W9-AVR-324

The Steel Industry in Japan

The Japanese steel industry has experienced both boom and recession over the past 30 years. It is now undergoing a thorough restructuring. Like the British steel industry, it has faced serious economic, technical and political changes. Harukiyo Hasegawa identifies, compares and analyses these changes, paying particular attention to the impact of modern technology upon all employees within the industry, from blue- to white-collar workers and managerial employees.

Local conditions, such as the difference in the ownership of the industries in the two countries, have led to different corporate strategies and, consequently, to the implementation of different technologies within the same technical paradigm. Hasegawa describes how, when and for what reasons particular technical innovations have been introduced, and examines the way in which management and labour have adjusted in the two countries.

Challenging the simplistic notion of 'leader' and 'follower' industries, Hasegawa introduces two important new conceptual tools for comparative study: 'relative advance' and 'convergence'. By bringing the notions of 'relative advance' and 'convergence' to bear on these case studies, Hasegawa is able to analyse whether the steel industries and, by extension, other manufacturing industries, can survive in mature economies.

Harukiyo Hasegawa is Lecturer in Japanese Studies at the University of Sheffield.

Sheffield Centre for Japanese Studies/Routledge Series

Series editor: Glenn D. Hook, Professor of Japanese Studies, University of Sheffield

This series, publishing by Routledge in association with the Centre for Japanese Studies at the University of Sheffield, will make available both original research on a wide range of subjects dealing with Japan and will provide introductory overviews of key topics in Japanese studies.

The Internationalization of Japan
Edited by Glenn D. Hook and Michael Weiner

Race and Migration in Imperial Japan
Michael Weiner

Greater China and Japan
Prospects for an economic partnership in East Asia
Robert Taylor

Race, Resistance and the Ainu of Japan
Richard Siddle

The Steel Industry in Japan

A comparison with Britain

Harukiyu Hasegawa

London and New York

First published 1996
by Routledge
2 Park Square, Milton Park, Abingdon, Oxon, OX14 4RN

Simultaneously published in the USA and Canada
by Routledge
270 Madison Ave, New York NY 10016

Transferred to Digital Printing 2005

Routledge is an International Thomson Publishing company

Typeset in Times by
Ponting–Green Publishing Services, Chesham, Bucks

British Library Cataloguing in Publication Data
A catalogue record for this book is available from the British Library.

Library of Congress Cataloguing in Publication Data
A catalogue record for this book has been requested.

ISBN 0–415–10386–X

To my parents

Contents

Part IV Convergence and relative advance

Figures

Tables

Acknowledgements

The rapid growth of the steel industry in Japan has, in my view, long required a compact and comparative assessment, and I am delighted to have been able to complete such an overdue project. It could not, however, have been realised without the kind assistance of a number of people from both the academic and industrial fields on both sides of the world. It would be impossible to name everyone to whom thanks are due, but there are some people without whose assistance the realisation of my research project would have been overwhelming.

In Japan, many at Tekkō Renmei, Tekkō Rōren, Nippon Steel Corporation, Kobe Steel Ltd, Nisshin Steel and Nihon Kōkan KK gave freely of their interest and information, for which I offer my most sincere thanks.

No less gratitude can be extended to those people in Britain who have made the study possible. Professors Charles Rowley, Charles Hanson, James Foreman-Peck, Ronald Dore and Peter Drysdale gave valuable advice and direction when I was working on my doctoral thesis, on which the present volume is based. Several individuals at the BSC, both at head office and works levels, and the ISTC provided valuable information and also clarified certain relevant issues for me.

The initial suggestion that I should publish my research came from Professor Glenn Hook, whose subsequent warm and constant encouragement made it possible for my work to appear in the Sheffield–Routledge Japanese Studies series. Professor Ian Gow generously allowed me study leave to work on the manuscript, while Professor Andrew Tylecote and Dr Robert Taylor have read some of the chapters and given helpful comments on the conceptual terms I have used. Follow-up research on recent developments in the two steel industries was greatly facilitated by the Japan Foundation.

Throughout my academic career, I have esteemed certain people as long-term mentors and friends, giving me intellectual stimulation and support, and I welcome the chance to thank Professors Toshitake Noma, Edith Shiffert, Masaki Nakata, Takeshi Watanabe and Kazuo Ishida in print.

I hope that this book will make an original and worthwhile contribution to the field I have chosen, but it is of course only a small piece in a substantial whole to which many other scholars have already contributed. These scholars should share any approval this work may receive, but responsibility for errors or other failings is entirely mine.

Finally, my thanks go to John Billingsley for giving me the benefit both of his editing skills and his knowledge of Japanese and English in tidying up and improving my original English; to my publishers, Routledge, and especially Gordon Smith and Victoria Smith; and to my family, particularly Michiko and Harutomo, who have supported and aided my long hours at the computer screen.

Abbreviations

AGC	(Automatic Gauge Control)
APACS	(Association of Patternmakers and Allied Craftspersons)
ASBSBSW	(Amalgamated Society of Boilermakers, Shipwrights, Blacksmiths and Structural Workers)
AUEW–TASS	(Amalgamated Union of Engineering Workers–Technical Administrative and Supervisory Section)
AUEW–E	(Amalgamated Union of Engineering Workers–Engineering Section)
AUEW–F	(Amalgamated Union of Engineering Workers–Foundry Section)
BF	(Blast Furnace)
BISRA	(British Iron and Steel Research Association)
BRTTS	(British Roll Turners' Trade Society)
BSC	(British Steel Corporation)
CSO	(Central Statistical Office)
DTI	(Department of Trade and Industry)
ECSC	(European Coal and Steel Community)
EEC	(European Economic Community)
EETPU	(Electrical, Electronic, Telecommunication and Plumbing Union)
GDP	(Gross Domestic Product)
GMBATU	(General, Municipal, Boilermakers and Allied Trades Union)
GNP	(Gross National Product)
HR	(Human Relations)
IE	(Industrial Engineering)
IISI	(International Iron and Steel Institute)
ILO	(International Labour Organisation)
ISTC	(Iron and Steel Trades Confederation)
LD	(Linzer–Dusenverfahren converter)
LSB	(Local Lump Sum Bonus)
MITI	(Ministry of International Trade and Industry)
MTP	(Management Training Programme)
NCCC	(National Craftsmen's Co-ordinating Committee)
NEBSS	(National Examinations Board for Supervisory Studies)
NJC	(National Joint Council)

NUSMWCH & DE	(National Union of Sheet Metal Workers, Coppersmiths, Heating and Domestic Engineers)
PR	(Public Relations)
QC	(Quality Control)
R & D	(Research and Development)
SCNI	(Select Committee on Nationalised Industries)
SIMA	(Steel Industry Management Association)
SMS	(Sub-management System)
T&GWU	(Transport and General Workers' Union)
TUC	(Trade Union Congress)
TUCSICC	(Trade Union Congress Steel Industry Consultative Committee)
TWI	(Training Within Industry)
UCATT	(Union of Construction, Allied Trades and Technicians)
UK ISSB	(United Kingdom Iron and Steel Statistics Bureau)
UK	(United Kingdom)
US	(United States)
USA	(United States of America)
USSR	(Union of Soviet Socialist Republics)
ZD	(Zero Defect)

Glossary

Buchō	(general manager or departmental manager)
Chōsa Geppō	(monthly survey)
Fukushachō	(vice president)
honkō	(regular worker)
Jiba Kigyō	(local firms)
Jishukanri Katsudō	(self-management activity)
Jyōmu Kai	(managing directors' council)
Jyōmu	(managing director)
Kagaku gijyutsuchō	(Science and Technology Agency)
Kakarichō	(section chief)
Keiei Kaigi	(Corporate Council)
Keieihōshin Kaigi	(Corporate Policy Council)
Keiretsu Kigyō	(affiliated enterprises)
Keizai Kikakuchō	(Economic Planning Agency)
Nihon Seisansei Honbu	(Japan Productivity Centre)
Nōryoku Kaihatsubu	(faculty development department)
Nōryokushugi Kanri	(management by ability)
Odan Kigyō	(horizontal firms)
Rōdō Handbook	(Labour Handbook)
Sanji-ho	(candidate for councillor)
Senka	(course for professional knowledge)
Shachō	(president)
Shitaukekō	(sub-contract workers)
Shitsuchō	(room chief)
Shokumukyu Seido	(pay system according to job function)
Shokunō Shikaku Seido	(job qualification system)
Shuji	(superintendent)
Shu tantō	(person in charge)

Sohyō	(General Council of Japanese Trade Unions)
Sōrifu Tōkeikyoku	(Statistics Bureau of Prime Minister's Office)
Tantō-ho	(candidate for person in charge)
Tekkō Kaihō	(Report of Steel World)
Tekkō Kyōkai	(Iron and Steel Institute of Japan)
Tekkō Renmei	(Japan Iron and Steel Federation)
Tekkō Rōren	(Japanese Federation of Iron and Steel Workers' Unions)
Tokatsu Shuji	(chief superintendent)
Tokei Yōran	(Handbook for Iron and Steel Statistics)
Torishimariyaku	(director)
Tsūsanshō	(The Ministry of International Trade and Industry)
Yukashōken Hōkokusho	(Annual Report and Accounts)
Zaibatsu	(financial conglomerates)

1 Introduction

This book examines continuity and change in the Japanese steel industry over the past thirty years in contrast to the steel industry in Britain. Our main concern is the impact of technology upon management and labour.

THE STEEL INDUSTRY

The heavy industrialisation of post-war Japan cannot be understood without taking into account the crucial role of the steel industry. Although numerous studies of the steel industry have been carried out in the framework of 'industrial studies' in general (Cockerill 1974; Ishida 1981; Okishio and Ishida 1981; Barnett and Schorsch 1983; Hogan 1983; Toda 1984; Hudson and Sadler 1989; ILO 1992), no study has yet been carried out on the steel industry over the past thirty years with a view to both management and labour and in comparison with the contrasting case of Britain. This study aims to fill this gap. But why is the steel industry appropriate as a case study in the context of management and labour?

The following reasons may be given. First, technological innovation in the post-war Japanese steel industry was extraordinarily rapid and extensive and its impact upon management and labour can be regarded as a typical case of change in management and labour. It was the steel industry which, ahead of other industries, adopted both technological innovation and management methods from abroad and reformulated them into a new production and management system. In addition, the end of expansion in 1974 and subsequent stagnation signalled a relative decline in this industry as well as the termination of rapid growth in Japan's manufacturing industry overall. The thirty-year period from 1960–90 is therefore indeed appropriate for an analysis of management and labour to examine its formation and reforms. Second, while the

Japanese steel industry experienced phases of institutional formation over the period 1960–73, and readjustment and restructuring from 1974 into the 1990s, the British steel industry has exhibited certain institutional reforms relevant to the production system within a framework of large-scale corporate restructuring throughout the period of this study. These contrasting processes exhibited in Japan and the UK allow us the opportunity to consider them within a proposed conceptual framework of relative 'advance' and 'convergence'.

The study of Japanese management and labour and its comparison with an equivalent industry in Britain have until now been limited to the electrical and automobile industries (Dore 1973; White and Trevor 1983; Wickens 1987; Garrahan and Stewart 1992), with no significant attention being paid to steel in spite of its importance among manufacturing industries.[1] Steel has long been regarded as a typical 'declining industry' in the UK, and presumably a comparative study with Japan has not been seen as of great value; British attention has been directed more towards newer industries such as the chemical, automobile, electronic and information industries. Another possible reason for the lack of attention is that there had been no direct investment of Japanese steel companies in the UK, although some steel companies had invested in the USA in the form of joint ventures after Japanese automobile companies began production in that country.[2]

In spite of this situation in the UK, the importance of the steel industry to both the Japanese and the UK manufacturing industries cannot be overlooked. For Japan the termination of the rapid growth and maturity of her economy needs to be considered more objectively from a macro-industrial perspective and for the UK the efforts to revive the manufacturing industry may be conceived in the context of the revival of a material supply industry such as steel. For whatever size and kind of manufacturing industry we need in the period of post-industrialisation, it is inconceivable that an innovative and sustainable manufacturing industry can exist without a successful restructuring of the steel industry. A comparative review of management and labour from a historical perspective would give us knowledge with which to create a future paradigm for both steel industries, most urgently for Britain and then for Japan.

Until recently, Japan's steel industry was ubiquitous. When a number of new large-scale steelworks came on line in the 1960s, rapidly expanding the industry's production capacity, no one thought or even dreamed that the Japanese steel industry might some day begin to decline. Employees in the major steel works were told and believed that their companies would continue to expand.[3] However, decline was not long in coming. In 1973, crude steel output reached 113 million tonnes but was never to approach that figure again. Employment in the industry, meanwhile, had peaked in 1970 but declined rapidly thereafter. We can say that within thirty years of starting its rapid revival and development following the Second World War, the Japanese steel industry went from rise to maturity and thence to relative decline.[4] Japan was not the only country to experience this rise and fall. The British steel industry, whose position in 1870 had been at an optimum position of relative 'advance', accounting for 49.8 per cent of world steel production, also declined, and its relative position came down even further in the post-war years. In the period of this study, the UK steel industry maximised its steel output at 28.3 million tonnes in 1970 and then shifted to a rather sharp decline.

Figure 1.1 shows such a shift in production and employment during the period under investigation. The general trend is that the production of steel has a pattern of rise and decline, while that of employment shows a constant decline during the period of this study, except for some increase in Japan during 1960–5 and towards 1990. This study assumes that during the production growth phase in Japan, which reflected a period of institutional formation, the institutional arrangements developed to a point which made people think that it was the institutional arrangements themselves that were the cause of Japanese business success. However, the decline since 1974, which reflects the beginning

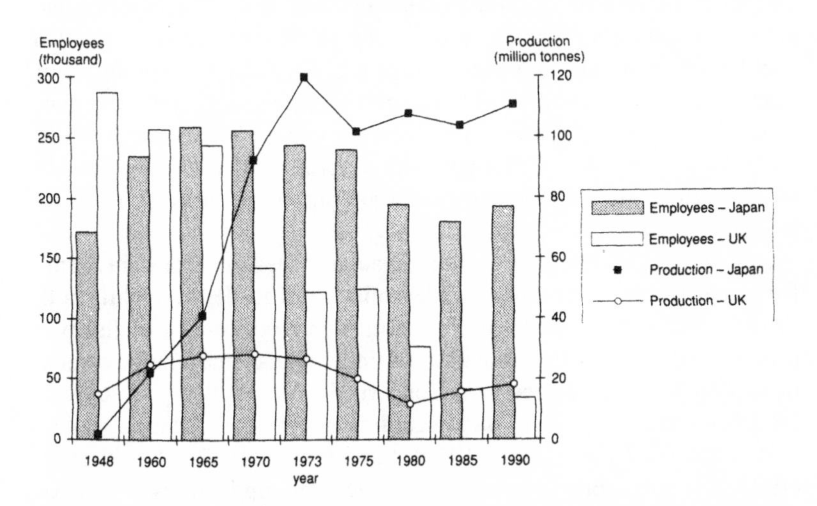

Figure 1.1 Steel production and number of employees in Japan and the UK
Sources: Tekkō Renmei (Tekkō Tōkei Yōran) 1970, 1980, 1985, 1993; Department of Employment and Productivity 1971; Iron and Steel Statistics Bureau (1980, 85)

of the phase of contrived management, has forced management to restructure first its institutional arrangements and then its corporate structure. In the UK's case, apart from a slight improvement around the mid-1960s, the whole period reflects a trend of gradual decline in both production and employment. The growth phase of Japan formed a production system which induced a convergence tendency in Britain in the late 1970s through the 1980s, while the ongoing process of restructuring in Japan's institutional and corporate structure from the latter half of the 1980s into the 1990s shows signs of Japan converging towards Britain. This, moreover, is encouraged as a response to the changing corporate environment in Japan and abroad.

TECHNOLOGY

Technology constantly influences the dynamics of management and labour in both factory and office environments. The changes that technology brings to the workplace generally imply the pursuit of higher productivity. This is a process many find difficult to resist, as it will bring long-term benefits to society as a whole; nonetheless, the short-term consequences of technological adaptation frequently conflict with the interests of employees. Within organisations, such a situation provokes a complex of responses – although some employees will benefit from, and thus accept, a new technological application, there are others who will not benefit, and will oppose change. In due course social relationships and values in an enterprise may also change, which in turn facilitate change in work practices, employment practices, working conditions, industrial relations and even the organisation's corporate structure. The increase and globalisation of world trade, as well as intensifying competitive pressures, encourage steady technological innovation, and hence alterations in management and labour.

Research on such technological impact has been an important theme for management practitioners, trades unions and academics; while both management and unions have been motivated by the desire to discover prescriptive answers to immediate business management policy questions, academics have been more concerned with enhancing our objective (long-term and overall) understanding of the phenomenon.

Research by Hunter *et al.* (1970), Forslin *et al.* (1979), Ozaki *et al.* (1992) and others has demonstrated the general implications of technological innovation for employment, work content, working conditions and management–labour relations.[5] This study, which aims to further elucidate the theme shared by these studies, compares the cases of

Japanese and British steel industries. A historical as well as comparative perspective is useful to examine the overall effects of technological change upon management and labour.[6]

In the discussion of technology and its impact, a major point concerns technological determinism, which implies a degree of emphasis on technological influence. Although it has been difficult to measure this quantitatively, a majority of academics recognise the general effect of technology *per se* and this is often discussed at the macro level around the theme of convergence in industrial societies (Marx 1965: 8–9; Kerr *et al.* 1960; Galbraith 1978: 403–6; Kerr 1983: 1–7).[7] The general hypothesis in such cases is that 'regardless of the initial institutional structure of an economy, the progress of industrialisation will make different systems of industrial relations and organisation converge' (Foreman-Peck and Hasegawa 1989: 29). A common technology will generate common social consequences.

The convergence theories of industrialisation have attempted to examine industrialisation and its influence upon social and economic changes. But such change was a macro-societal consequence of micro-corporate change. Thus, 'convergence begins with modern large-scale production, with heavy requirements of capital, sophisticated technology and, as a prime consequence, elaborate organisation' (Galbraith 1978: 403).[8] Braverman (1974), arguing from a Marxist perspective, demonstrated how, in the long term, technology influences and then consequentially degrades labour by the separation of conception from execution. Of late, this emphasis on technological determinism and its neglect of working class consciousness, organisation and activities has been criticised.[9] Even so, no approach taking account of periods of ten or twenty years can deny the influence of technology (Armstrong 1988). The view of technology held in this work is relative as well as long term – relative, because we see it as a response of management as well as labour to the various factors of their corporate environments over the period of about thirty years. Thus, for example, the nature of employment differs from country to country, and even company to company when examined in detail, reflecting the dynamics of management and labour in their history – yet in spite of such differences, there remains room for 'convergence' in the long term via a similar technological condition[10] adopted to increase corporate profitability.

The macro-convergence issues of the 1970s and early 1980s were joined by a discussion of Japanisation and Japanese-style management in the late 1980s, adding a new dimension to contemporary management and Japanese studies. In fact, the essence of this topic concerns convergence at the micro-corporate level, as attempted by Dore (1973).

The themes found in the Japanisation discussion are whether Japanese-style management is transferable, to what extent this may be possible, what may prevent such transfer, what the consequences may be and whether such transfer is desirable or not (Dunning 1986; Wickens 1987; Oliver and Wilkinson 1988, 1992; Garrahan and Stewart 1992; Elger and Smith 1994). These transference discussions imply that Japanese-style management, which is based upon a new technological circumstance, can become either explicitly or implicitly an object of convergence.[11] Managerial efforts to emulate a relatively advanced technological condition can thus be interpreted as a force for convergence. Although Japanisation became widely known in the late 1980s, when Japanese direct investment in Britain increased dramatically, some moves towards emulation had already, by the late 1970s, been made by the British steel industry (BSC/TUCSICC 1975).[12]

The hypothesis in this research is that if any evidence exists of such convergence at a macro/societal level, then it can also be identified at the micro/corporate level. The objective of this study therefore is to examine the process of and explain the consequences of corporate level convergence using the conceptual framework of relative 'advance' and 'convergence' introduced in this chapter. This study will specifically consider changes in management and labour, focusing upon the respective technological innovation phases in the two steel industries under review.

APPROACH

The conceptual framework of relative 'advance' and 'convergence' used in this study is a guide and tool for the analysis and interpretation of changing phenomena in management and labour. Relative *advance* describes a state of comparative advantage in production system or corporate structure. It becomes an incitement to emulation by others, which manifests itself as a phenomenon of *convergence*. In this study the Japanese steel industry is investigated as a typical case of relative advance in production system and that of Britain as an advanced form of corporate structure. Thus is assumed the possibility for two kinds of convergence, one a British convergence towards Japan and the other a Japanese convergence towards Britain.

The first point for consideration is the relative advance which derives from production systems. It emerges from production technology, work practices and working conditions, which together constitute a production system. The production system dealt with here is part of a corporate management system, which is itself composed of three sub-management systems (SMS). The management system, in which the production system is included, interacts with three corporate environments: political/legislative, socio-cultural and economic. Figure 1.2 shows how these elements of management are systematically interrelated. Mutual compatibility is assumed to be a precondition for any relative advance created by production systems and implies some degree of non-conflictual and harmonious functionality obtaining between the sub-systems of management and production in a given corporate environment.

Technology at the corporate level is a key factor in the production system and can be examined by reference to two associated processes, its introduction, and its impact upon management and labour.

The first process concerns the corporate environment, in particular the influence of economic factors. As a given macro-economic condition, they are directly relevant to corporate growth,[13] which is usually achieved through technological innovation. Other factors are also

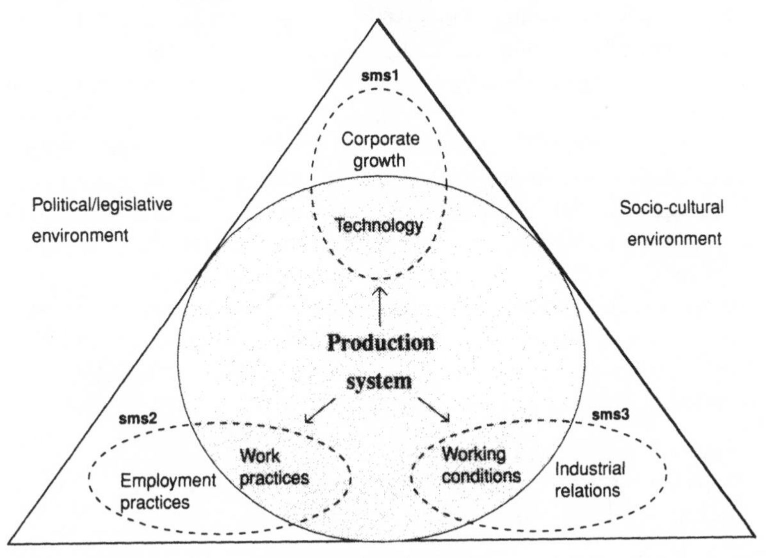

Figure 1.2 Compatibility in management systems

Notes (a) The production system consists of technology, working conditions and work practices

(b) The management system consists of three sub-management systems: sms1, corporate growth and technology; sms2, industrial relations and working conditions; sms3, employment practices and work practices

(c) The corporate environment consists of: the political, legislative factor, the socio-cultural factor, and the economic factor

important, but to a degree which varies according to the state of social and economic development in a particular industrial society. The phase of the economic cycle may also become a determining factor.

The second process concerns how technology is employed and its impact on both management and labour. Technology in this study is understood as a concept which includes both the technology itself and the methods of its effective employment, which will thus influence work practices and working conditions.

Any production system includes three factors of management: technology, work practices and working conditions. The production system thus defined is currently accompanied by the extensive use of automation. But the area and level of automation is constrained by both technical and economic principles.[14] This is reflected in the differing levels of automation from factory to factory and even workplace to workplace. Some areas therefore still remain labour-intensive, even if automation is technically possible. Nonetheless, in general the main production processes are equipped with sophisticated computers; we can therefore take production systems to be at an advanced stage of automation. Such conditions induce management to create a suitable form of division of labour and social relations, manifesting itself in work practices and working conditions and representing an incorporation of the technical system into the social system of management.

Management, when not used in the sense of an organisational layer of managers, is defined in this study as the function of managers to *integrate* and *control* various managerial areas and objects, in which we include labour, in order to achieve both macro and micro corporate objectives such as profitability, growth and market share as well as cost, quality and stability. The managerial areas and objects dealt with in our study can be classified into the following three sub-systems, as encircled in Figure 1.2:

1 corporate growth and technology
2 employment practices and work practices
3 industrial relations and working conditions.

The first sub-management system functions to establish an important part of the production system, while the second and third will usually be made to correspond to the first. The second and third sub-systems will interact with each other, resulting in a certain degree of compatibility between them. For the whole system to be coherent, rendering the corporate organisation effective and efficient, the factors in these sub-systems require compatibility both among themselves and in the inter-relationships between them.

The influence of corporate environments is considered in two stages. The first is a stage favourable to corporate activities because they are more or less in the process of formation as heavy industrialisation proceeds. In this phase the nature of compatibility is organic – more natural and tending towards growth. The other stage is one in which industrial growth has almost come to a halt and there emerges a tension between the corporate environment and the management system. This is a time when firms are forced to readjust in order to overcome the resulting difficulties, thus shifting the nature of compatibility from *organic* to *contrived.*

While these six factors are the objects and areas of management, those involved in the business organisation are employees of various kinds: blue-collar, clerical, technical and managerial. System coherence therefore implies that at any time when a certain degree of compatibility is maintained, technical as well as social control is exerted to induce the identification of employees with corporate objectives. This control is exercised through various rules and practices, though these may differ according to the business environment. In this connection, any education and training within the organisation, which ideally aims to develop human resources, may become a specific agent for both control and integration when the industry and firms are growing, as was the case in Japan.[15]

Labour is defined rather broadly in this study and is considered in both its quantitative and qualitative aspects. The former is simply the total number of employees and their composition at the various levels of the business hierarchy; such changes are a response to the dynamics of the management system. On the other hand, the qualitative aspect is the combination of skills and knowledge internalised in each employee, and their feedback to management. Employees possessing certain attributes in this regard are officially classified into occupations and jobs and further divided into grades, although Japan and the UK differ in their terms of classification. Human labour is thus technically as well as socially controlled by various rules and courses of action which manifest themselves as employment and work practices. This process of control forms a characteristic of social relations in the business organisation.

In effect, employees work under prescribed work practices and working conditions, while at corporate level they are under employment practices and industrial relations which mainly result from the interplay of management and labour/trade unions. Feedback within this interplay, whether actual or potential, may occur as resistance whenever technical and social control devalues labour in either or both the economic and social/human areas – most specifically, in wages/salaries and immediate working conditions. Such resistance to managerial control colours the nature of industrial relations which are a major part of social relations in any business organisation: with industrial relations conforming to one of the following four types – compromise, conflictual, co-operative or compliant – we see a shift in their nature in both steel industries. If the use of technology in an enterprise is perceived by workers to be designed or utilised in such a way as to entail the alienation or degradation of labour, worker opposition – which may lead to union action – constitutes a social process. Therefore, production systems, in which technology is a vital factor in how labour is accommodated, tend to differ from country to country, and even from company to company, depending upon the stage of industrialisation and other individual factors of management, particularly the power relations manifest in industrial relations.

Analysis of workforce composition is extensively used in this study in order to examine the nature of such change in management and labour. Data on workforce composition will be given for the iron and steel industry across a wide range from blue-collar employees to top management. Workforce composition is seen here as a reflection of the division of labour established within given technological conditions. In the extant literature dealing with labour, the scope of research has often been limited mainly to production workers, with much less attention paid to clerical, technical and managerial employees (Ishida 1981). In addition, sub-contract workers (a phenomenon mainly found in Japan) have usually been excluded from analysis as they are under the management of sub-contracting companies.[16] In this study, sub-contract workers are treated as an integral part of the main labour force and as an object of managerial control under a common production system.

Second, relative advance in corporate structure can be discussed in terms of a historical perspective of corporate development. Relative advance in corporate structure is a consequence of deliberate managerial efforts to cope with changing aspects of the corporate environment, in particular that of economic influences. Termination of rapid economic growth induces reorganisation of the corporate structure. Increased productivity due to advanced production systems and the ending of market growth makes an enlarged functional organisation unnecessary and requires management to restructure it, particularly to shift the relative importance of corporate objectives from growth to profitability by reducing the size of their main business and creating clear product-based divisions. They need to fill the gap between their specific production capacity and the needs of the market. This suggests a termination of organic growth and the beginning of contrived survival. The case of the British steel industry is typical of this experience, and implies a recent Japanese convergence towards Britain. Any move in the nature of compatibility in management implies a shift in managerial principles of employment practices and industrial relations, which tends to create difficulty and conflict in industrial relations. Relative advance in corporate structure is not therefore an incitement to 'favourable' emulation, but rather a necessity that must be adopted in order to survive.

The above framework for explaining the phenomenon of convergence is a perspective emphasising the existence of universal trends in the way business organisation develops, although it does not imply that convergence tends to make management and labour identical in the two countries, diminishing individual characteristics. Due to the various differences in corporate environments, as shown in Figure 1.2, perfect convergence is highly unlikely even where similar corporate objectives and use of production technology exert comparable forces upon the dynamics of management and labour.

Our hypothesis is therefore that in Britain a process of convergence towards Japan has occurred in the area of production systems, because Japan succeeded in establishing its advanced production systems before Britain. On the other hand, a process of convergence towards Britain in the area of corporate structure is occurring in Japan, due to the termination of rapid growth there in the mid-1970s. Britain displays a mature form of corporate structure, deriving from a mature economy. The hypothesis of the convergence of Britain towards Japan can be illustrated as in Figure 1.3.

Figure 1.3 shows how British convergence occurs due to relative advance in the production systems of Japan. Relative advance is assumed to occur when the factors of the X and Y axes, which are explained as factors of management in Figure 1.2, become compatible. Optimal relative advance in production systems will appear along the 45° line and denotes where individual corporate compatibility, together with the rise of the industry in which the firm belongs, can be achieved. The X axis includes corporate growth and technology, while the Y axis includes factors of institutional arrangements such as employment practices, work practices, industrial relations and working conditions. As the line of optimal relative advance is approached, a higher level of compatibility occurs, mainly on the basis of innovation in production systems. A convergence in the area of production technology will therefore occur towards a country which has peaked further along the line of relative advance.

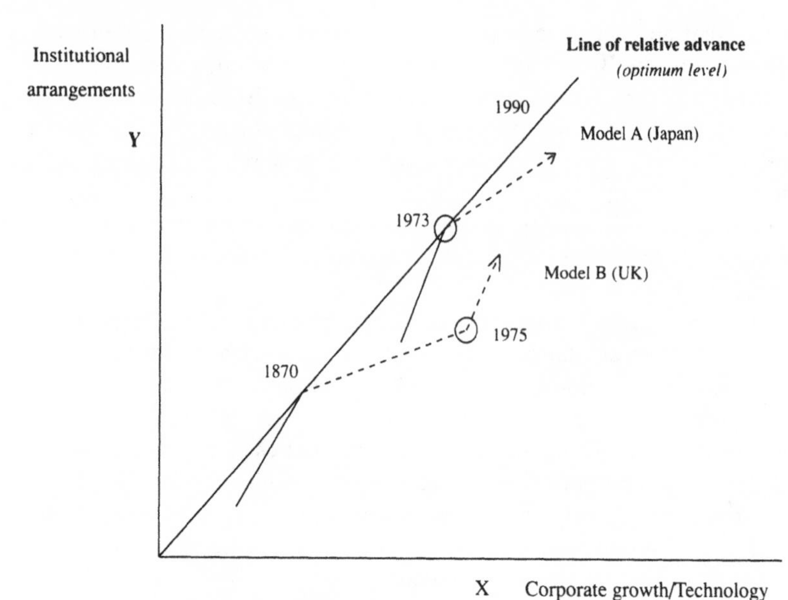

Figure 1.3 British convergence towards Japan: relative advance in production systems

Notes: (a) Institutional arrangements include work practices, employment practices, working conditions and industrial relations

(b) ______ implies a stage of organic compatibility, while _ _ _ _ _ _ _ _ indicates that of contrived compatibility. The term 'organic' indicates a natural process of institutional formation, while the term 'contrived' indicates a struggling process of re-adjustment in which serious managerial efforts are made to maintain compatibility in management

On the other hand, a convergence in the area of corporate structure will emerge when an industry passes its point of optimal advance and begins to depart from the line; at this stage, movement will occur towards an industry which has already experienced and dealt with this decline. Britain, which passed its peak a considerable time ago, and now exhibits an 'advanced' form of corporate structure, will thus be the direction in which convergence from Japan is likely. This should not be seen to imply perfect convergence at some future point, as the constant flux in variables such as factors in corporate environment may invite a shift in trends in response to changing circumstances. Nevertheless, in terms of the period under study here, the dynamics of convergence shown in Figure 1.4 can be demonstrated; British convergence in the area of production and management towards the Japanese model is diagrammatically represented as a move along the X

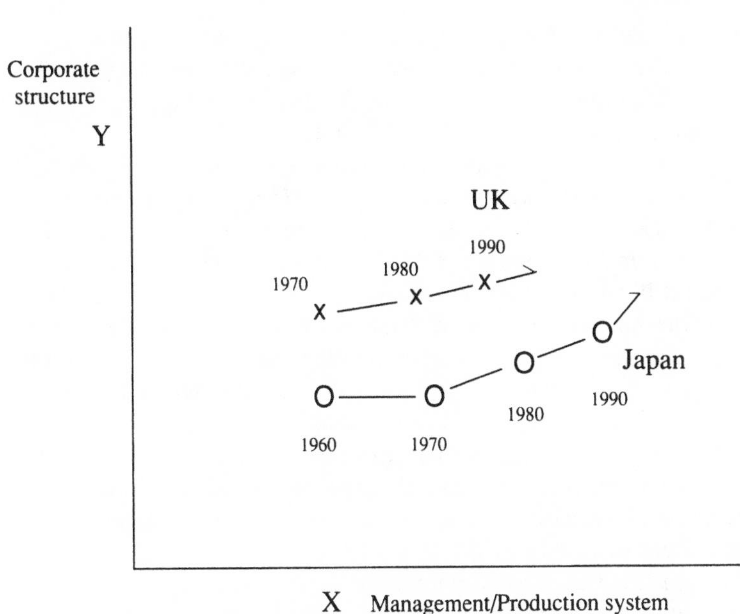

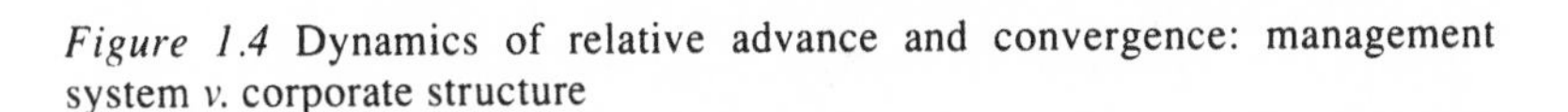
Figure 1.4 Dynamics of relative advance and convergence: management system *v.* corporate structure

axis (showing a time lag factor of 10–15 years), while Japanese convergence towards Britain manifests as an upward shift along the Y axis, taking place following the end of its rapid growth and throughout the 1980s and 1990s. The convergent trend can be seen easily.

Two types of compatibility in management can be identified and are characterised in this analysis as 'organic' and 'contrived' compatibility.

Organic compatibility implies a state and process whereby institutional arrangements – such as work practices, employment practices and industrial relations – can be formed naturally and unconsciously as a system of management, i.e. with little deliberate effort. In Japan's case, these conditions emerged between 1960 and 1973 without a great deal of controversy, seeming simply to follow on from the large-scale technological innovations underway at the time (albeit in a manner influenced by cultural predisposition). In the case of Britain, despite managerial efforts, return to the line of relative advance in production systems has been elusive and therefore an optimum level of system compatibility has not been regained. However, managerial efforts to emulate the advanced production system achieved in Japan have spurred some institutional reform relevant to work practices and industrial relations. A degree of compatibility has thereby been achieved and has aided organisational efficiency in production systems as already shown by a modification of the Model B (UK) line in Figure 1.3 and as a movement along the X axis in Figure 1.4.

In general, we can assume that a relative advance in a production system is realised through the process of applying a number of technological opportunities, either from within or without, at a time when the national economy is achieving full-scale industrialisation. As Figure 1.3 illustrates, the UK achieved these conditions[17] in the latter half of the nineteenth century,[18] and the steel industry in particular reached its peak around 1870. On the other hand, Japan reached this stage by utilising opportunities mainly from without in the period 1960–73,[19] and then departed from its peak of compatibility. While the UK steel industry was making huge investments in the latter half of the 1970s, some convergence towards Japan occurred, and this can be understood as an early sign of the *Japanisation* which appeared and spread more distinctly in the late 1980s among other British manufacturing industries. Thus Japanisation has become a reflection of the convergence phenomenon of Britain towards Japan in the area of production systems.

Contrived compatibility implies a state and process of managerial efforts to restructure corporate organisation in order to find a way to survive the end of growth and the onset of de-industrialisation. Institutional arrangements are reviewed in order to realise as high a degree of compatibility as possible, or to halt further departure from the line of relative advance. It will invariably require conscious effort and create various difficulties of adjustment, and it results in a profit-oriented division system (divisionalised organisation) in the corporate structure. The Japanese steel industry departed from the optimal compatibility line from around 1974 and suffered uncertainty in the next decade. In the late 1980s it began to show signs of convergence towards Britain in the area of corporate structure in order to cope with the decline in profitability and stem the loss of relative advance (Figure 1.4). This proceeded further in the 1990s, with Japan's continued recession. Japan has thus begun to show signs of convergence towards Britain as the situation of contrived compatibility develops, in the process of reform in employment practices as well as corporate structure. This convergence of Japan towards Britain has been accentuated as the steel industries in Korea and China have achieved their own technological innovations and become increasingly successful rivals. Although institutional reforms in Japan will continue, difficulty may be expected in returning to the original optimum point of compatibility.

THE STRUCTURE OF THE BOOK

This book is composed of four parts. Part One (Chapters 2, 3 and 4) is a comparative section examining the three areas of corporate environment for the steel industry. Chapter 2 reviews the macro-economic factors relevant to corporate growth. We compare aspects of the economic, industrial and institutional dimensions which we believe to be important in determining the timing and type of technological innovations implemented. In Chapter 3 we compare four dimensions of industrial organisation and examine them in terms of competitive incentives and pressures. We consider the implications of change in both countries through different processes, Japan through competition and Britain through nationalisation, towards oligopolistic conditions weakening the competitive pressures in the domestic market. Also examined are various aspects relevant to the above dimensions which have influenced the adoption of an advanced production system. Chapter 4 compares four institutional factors and considers their relevance to capital investment. The given institutional practices are reviewed as an internal corporate environment, although they are usually considered in the current approach to be a consequence of both managerial strategy and social relations within business organisations. Our assumption in this chapter is that within business organisations the institutional practices of today exert forces of inertia on tomorrow's decision-making.

In Part Two (Chapters 5, 6 and 7) the Japanese steel industry is examined, followed by that of Britain in Part Three (Chapters 8, 9 and 10). In each of these parts the process of technological innovation is reviewed; its characteristics are identified, and its impact upon management and labour investigated. Each part begins with an introductory chapter giving an overview analysis of the industry under review, and then proceeds to focus on an example case study of a steelworks considered broadly representative of that industry. These works are taken from those under the control of the main steel producers of each country in the period under study, i.e. the five major corporations of Japan (which produced on average more than 70 per cent of the total Japanese crude steel) and the BSC (which produced more than 85 per cent of the total British crude steel). These producers, though not the whole of their national industries, are so dominant in the scale of modernisation, production figures, employment and industrial trends that they are considered for the purposes of this research at least to count as representative of their industry in general.

The analysis is specifically centred upon the time when capital investment was concentrated in each industry, in order to highlight the impact of technology upon management and labour. Both industries seem to share a different timing and process of innovations, but have ended up with similar production systems with some variations in the direction of convergence. These reflect the interaction and dynamics of sub-management systems responding to their corporate environments.

In Chapter 5 attention is paid first to the process of rapid and large-scale modernisation in the Japanese steel industry, which resulted in the formation of a new production system. The process of establishing a new production system can be understood as a stage of organic compatibility in the context of our theoretical assumption of relative 'advance' and 'convergence'. In Chapter 6 we examine the change in management and labour at the works level, which occurred with the introduction of a new production system based upon a comprehensive on-line information system at Kimitsu works. The processes found in management and labour in this chapter may be construed as a management effort to create organic compatibility. The system known as *Nōryokushugi Kanri* (management by ability) is a reflection of managerial efforts to remain at a high level of compatibility despite the termination of organisational growth. Chapter 7 looks at management and labour in the head office in terms of the division of labour at head office and in-firm education and training. Head offices expanded as single-unit organisations without any significant diversification and the whole landscape of management and labour at head office, even after organisational growth had terminated, shows a situation of organic compatibility in management.

The British steel industry is then reviewed in Chapter 8, in the same way as the Japanese steel industry in earlier chapters. We examine the changes in management and labour which have occurred, unique adaptations based upon the already-established division of labour and social relations. These reflect a stage of contrived compatibility and the consequences of modernisation and rationalisation suggest the potential for a degree of convergence towards Japan in the use of the workforce at the workplace. Chapter 9 focuses upon changes in management and labour at works level and identifies differences from the Japanese experience. The examination in this chapter, in particular of works organisation, occupational divisions, sub-contracting and the role of the foreman will provide evidence to consider such changes in relation to the main theme of our study. Chapter 10 presents important evidence that the corporate structure and the function of the head office of the BSC reflect an advanced stage of corporate structure, and one based upon an alternative concept of profit orientation to that of Japan.

Industrial democracy in the form of employee directorship in the BSC did not seem to function as expected either for management or for the union.

A separate review of each industry in Parts Two and Three was deemed necessary as each presents certain difficulties in direct comparison, owing to differences in history and corporate development. For example, a substantial analysis of the sub-contracting and the education and training systems is included in Part Two, but these phenomena were far less important in Britain. Instead, shifting trends in wages and salaries, their relationship to occupations, and modifications in corporate structure due to the frequent turnover of top management are examined at length in Part Three.

Part Four (Chapters 11, 12 and 13) returns to a comparison of the important findings of the previous two sections, examining them in a framework of relative advance and convergence. Chapter 11 deals with a detailed comparison of Japan and Britain. The similarities and differences in management and labour are explained by various factors, with a new production system as a major factor for change. In comparison we can generalise that Japan's experience in the high-growth period was a case of organic compatibility in management, while Britain's was one of contrived compatibility in management. In Chapter 12 we attempt to classify various findings in terms of convergence. The findings are classified according to the direction of convergence: type 1: convergence of the UK towards Japan; type 2: convergence of Japan towards the UK; type 3: non-convergence.

Type 1 occurs through relative advance in the production system in Japan, while type 2 occurs due to relative advance in corporate structure in the UK. Non-convergence seems to occur because of marked differences in the corporate environment, particularly due to political, legislative and socio-cultural factors. These will also gradually change over time, but at any point in history factors of the corporate environment, in particular of culture, will appear as preconditions for managerial strategy.

The final chapter follows up changes in management and labour in the early 1990s and considers their implications in the light of the conclusions reached in Chapter 12. The changes in management and labour occurring in this period in both steel industries reflect management endeavours to maintain compatibility at the contrived stage, by vigorously pursuing the re-organisation of the corporate structure as well as the production system on the basis of this on-going corporate reconstruction. In this process the managerial prerogative was enhanced partly due to the drastic change in workforce composition, resulting in an altered climate of industrial relations towards conditions of union 'compliance'. Japan seems to have narrowed the gap with Britain in terms of the corporate environment, while maintaining a higher level of steel production than Britain.

Although the question of finance is an important one in terms of corporate strategy, it has not been considered in detail here, as the primary emphasis in this study is on the processes and implications of technological innovation. Chapters 5 and 8, however, deal respectively with the volume and source of investment in Japan and the volume and political significance of investment in the BSC. Finance has been taken to follow technological innovation, with management assuming that the reward from such innovation will offset investment costs. From the current perspective, therefore, it constitutes part of the objectives of the firm and detailed comparative analysis was not deemed appropriate for the objectives of this study.

Part I

Differences in corporate environments

2 Economic development

This chapter is primarily concerned with looking at macro-economic factors relevant to the development of the steel industry in Japan and the United Kingdom. The topics under examination will be the manner in which the steel industries of the two countries have been modified by factors such as steel intensity, industrial policy, labour market and import/export arrangements and the implications of this for their respective processes of technological innovation.

STEEL INTENSITY[1]

Any relatively sudden and major industrialisation creates a huge increase in the demand for steel. We can as a consequence assume a certain relationship between stages in industrialisation and the development of the steel industry, and in particular between the level and growth rate of the GNP and steel demand. This demand is a vital prerequisite for growth and technological innovation in a steel company. Research carried out by the International Iron and Steel Institute (IISI 1972) and a related report made by the International Labour Organisation (ILO 1992) set out the relationship between *per capita* GNP and steel consumption behaviour, which is commonly expressed by the term 'steel intensity':

1 A significant growth in steel intensity does not occur until income reaches $300 (1963 prices) per head, the minimum required before an economic take-off can be expected.
2 Thereafter, rapid industrialisation normally sets in, giving rise to a high rate of investment in production facilities and infra-structure which propels steel consumption upward faster than GNP growth, and thus results in a rapid rise of steel intensity.
3 In the next phase, industrialisation shows more balanced growth and steel intensity tends to stabilise.

4 Eventually, industry begins to extend into more highly sophisticated spheres and the service sector expands in relative importance. This phase normally occurs when average income reaches a level of around $2,500 (1963 prices); at which point steel intensity begins to decline.

(IISI 1972: 24)

Japanese *per capita* GNP was $622 in 1963 when the major steel companies began their full scale modernisation. They enjoyed a rapid rise of steel intensity. Britain, meanwhile, was already at the stabilisation stage with a *per capita* GNP of $1,602 (IISI 1972: 33, 53). Japan's GNP subsequently grew apace and with it the domestic demand for steel. Britain, however, did not experience such a large increase. Domestic demand for steel there remained at about 20 million tonnes of crude steel, rising only temporarily in 1965 and 1970, and in the latter half of the 1970s there were years when the level even dropped below that of the early 1960s.

In Japan, steel intensity, as expressed in Kg. crude steel consumption per US dollar of GDP at 1980 prices, increased dramatically from 0.078 to 0.104 in the period 1960–70, and then began to decline.[2] During the same period, the UK experienced continuous decline and in 1970 their intensity figure stood at 0.056. Thus the macro-economic environment, as reflected in steel intensity over the years under review, suggests that Japan was endowed with far more favourable conditions for a major modernisation of the steel industry than the UK. This study maintains that high steel intensity reflects a relatively early stage of industrialisation, with Britain in that sense being more advanced than Japan.[3]

The above trend of steel intensity is reflected in the contemporary rate of economic growth. First, comparing GNP growth rates between 1960–4, Japan's average was 12.0 per cent and Britain's 3.8 per cent. In 1965–9 the respective figures were 10.8 and 2.6 per cent; in 1970–4, 6.2 and 2.8 per cent; in 1975–9, 4.7 and 1.8 per cent. It can be seen that Britain's growth rate was consistently low in comparison to the extremely high level experienced by Japan. Both economies had considerably higher growth in the 1960s than the 1970s, implying that the former decade was for both the most vigorous period for capital investment.

In fixed capital formation, which basically conditions the nature of economic growth, the rate of gross fixed capital formation as a fraction of GNP in Japan was between 30–40 per cent during 1960–78, while it was only between 10–20 per cent in the UK. If one inspects the composition of gross fixed capital formation, the percentage of private sector plant and equipment investment and private sector housing investment were high in Japan; this is particularly true of the latter, which increased steadily after 1960 (from 4.8 per cent in 1960–5 to 7.1 per cent in 1971–8). In the UK, the story was once again of decline; from 1.9 per cent in 1960–5 to 1.7 per cent in 1971–8 in private sector housing investment. There is also a correlation indicating that the higher the percentage of gross fixed capital formation, especially in private sector plant and equipment investment, the higher the rate of economic growth. The proportion of private sector plant and equipment investment in Japan was 16–19 per cent over the whole period, while in the UK it was between 8–9 per cent.

The above levels of private sector housing investment and private sector plant and equipment investment are reflected in the domestic demand for steel. Comparing the two nations again, the delivery of ordinary steel products to the four major manufacturing industries and the construction industry, in terms of total deliveries, shows a four-fold increase in volume in Japan between 1960 and 1979; British deliveries decreased by 50 per cent in the same period. In Japan, the industries with the largest increase in the rate of deliveries were motor vehicle manufacturing and construction. These two alone accounted for over 30 per cent of total deliveries, each with more than six million tonnes in 1979. It is clear, then, that of the five industries whose demand for steel products soared, the motor vehicle and construction industries above all played the most important role in promoting plant and equipment modernisation in the Japanese iron and steel industry.

It is to be noted that in Japan the percentage shares of the mechanical engineering, electrical engineering and shipbuilding industries were declining. Shipbuilding had had a high share of deliveries up to 1975, but was then overtaken by the advance of the motor vehicle manufacturing industry, the most important sector after construction.

In the UK, the decrease in the amount of deliveries can be observed in every industry. Three show a rate of decrease of over 50 per cent. Among these, the shipbuilding and construction industries were especially hard hit, suggesting that the decline of these industrial activities was the major factor leading to the decline of the iron and steel industry. This is particularly important when we consider that in Japan their development, together with that of the motor vehicle manufacturing industry, was crucial for the expansion of the iron and steel industry. Industries with a lower rate of decrease were motor vehicle manufacturing and mechanical engineering. The percentage shares of these industries in Britain were among the highest in any period chosen for inspection; vehicle manufacturing, for example, accounted for about

50 per cent in 1979, indicating that it was this industry that was propping up iron and steel production in the UK.

Thus the growth of the Japanese construction, shipbuilding and motor vehicle manufacturing industries helped promote and maintain the burgeoning iron and steel industry by providing a massive demand for steel products. In the UK, it was motor vehicle manufacturing and mechanical engineering which had a determining influence on the prospects of the iron and steel industry, as the combined percentage of total deliveries accounted for by these two industries was as high as 71.5 per cent in 1979, showing a very high concentration which contrasts with the spread over a comparatively larger number of industries, as occurred in Japan.

This shows that the steel industry cannot grow in isolation, but involves the simultaneous growth of other steel-consuming industries, and depends strongly on the stage and process of industrialisation. The speed and level of development of these industries must in turn have influenced the timing, scale and form of steel modernisation. The scale and speed of Japanese steel modernisation was therefore a consequence of its relatively delayed large-scale industrialisation, which was itself predicated upon the political and economic conditions in the revival period of 1945–55.

INDUSTRIAL POLICY

Industrial policy can be regarded as a factor of economic influence, as shown in Figure 1.2, and it forms a part of the corporate environment which shapes corporate growth. From a political perspective it has been regarded in Japan as a vital tool for the development of the national economy. In the UK, on the other hand, its importance was relatively small and was moreover being replaced by direct political commitment in the form of nationalisation. In stark contrast to the UK, the Japanese government considered the iron and steel industry to be a vital factor in the national economy.

The nature and degree of commitment of the two governments towards their steel industries needs to be reviewed. Two Japanese government reports set a high value on the steel industry and declared its necessity not only for a major industrialising initiative, but also for expanding exports: these were the report concerning the 'New Long-term Economic Planning', published in 1957 by the Mining and Manufacturing Sector of the Government Economic Council; and the views on 'Liberalisation of the Iron and Steel Industry', published in

1960 by the Committee for Trade Liberalisation Measures of the Japan Iron and Steel Federation.

The first report states in its first chapter that the objective of planning is to achieve full employment and a steady rise in the living standards of the nation as a whole. For the fulfilment of the plan the production levels of mining and manufacturing in 1962 were to be increased by 60.5 per cent over those of 1956. In this connection it emphasises the necessity of installing new blast furnaces and LD converters. In addition, Chapter 5 of the report, which gives a summary of policy, introduces views on and aims for heavy industrialisation:

1 to strengthen international competitive strength and promote exports;
2 to achieve capital accumulation and secure important equipment funds;
3 to promote scientific technique and nurture new industries;
4 to consolidate conditions for industrial location;
5 to stabilise long-term demand and supply for basic raw materials;
6 to develop domestic resources and promote their efficient use;
7 to encourage business activities of small and medium enterprises.

Implicit in the first report's approach to post-war Japanese industrialisation was that all other measures were to support this basic industrial policy. The other document, which reflected the emerging corporate environment, noted that in competition with the advanced nations of the West, Japan suffers from the relatively high cost of raw materials and the high interest rate levied on borrowing, and it requested the government that the following measures should be implemented as soon as possible:

1 Countermeasures to strengthen international competitiveness of enterprises:

(a) To strengthen enterprises by giving priority to capital accumulation (taxation):
- to continue the special depreciation system
- to shorten the depreciation period
- to reduce corporation tax and enterprise tax
- to amend the system of taxation on dividends
- to make a Voluntary Depreciation System for R & D equipment and the Preparatory Fund System for Experimental Research.

(b) Measures for the funds:
- to ease the acquisition of funds by the introduction of foreign capital
- to lower interest rates.

(c) Measures for technology:
- to make the state carry out basic research
- to strengthen technical education
- to promote other and general technology
- to give favourable treatment to research carried out by enterprises.

2 Countermeasures for raw materials:

(a) To make the best use of the effects of liberalisation for raw materials:
- coal: to allow free use of cheap imported coal
- crude oil: to allow free imports of crude oil
- transport of raw materials: availability of free choice for the cheapest vessels in the world.

(b) To implement various measures in order to lower the price of domestic coal to levels comparable to those of the major competitive countries.

(c) To promote the development of overseas raw material and to establish countermeasures for transport, such as building exclusive vessels.

3 Countermeasures to maintain industrial order:

(a) To revise excessive competition and establish a co-operative system.

(b) To strengthen the Open Sales Scheme.

(c) To revise the Anti-monopoly Law.

4 Measures to protect from low price exports of other countries (revision of tariff):

(a) To revise the present *ad valorem* duties to specific duties depending upon production and necessity, and to make flexible use of this tariff.

These comprehensive and strict protective measures for the steel industry were almost all implemented during the 1960s, and greatly contributed to the establishment of steel as Japan's key industry, as well as a prime export industry.

By contrast, the main governmental industrial policy of the UK was the renationalisation of steel in 1967. It aimed to modernise and increase competitive strength by transferring ownership from private to public interests. The nationalised British Steel Corporation carried out large-scale rationalisation and modernisation in the 1970s, particularly in the latter half of the decade. Although this direct commitment in the form of ownership was made, financial aid constituted the bulk of other indirect industrial policy. We can see this in the report of the Federal Trade Commission (1977: 355–6) on the financial aid packages given during the period 1968–76, which discussed:

- possible interest reduction on public loans
- interest reductions on European Investment Bank loans
- non-payment of dividends on Public Dividend Capital
- interest and dividend savings associated with the write-off of public debt and Public Dividend Capital
- regional development grants and losses covered by government ownership.

The above financial aids in the form of subsidy value were calculated to be 11.14 pounds per metric tonne. In addition, the report from the Iron and Steel Trades Confederation (ISTC 1980) concerning financial aid from the government to the British Steel Corporation says that 'the BSC is denied so much of the state support given on the continent, that British steel is put at an unfair disadvantage' (ISTC 1980: 52). The ISTC insisted that if the following state subsidies which were received by its European rivals from their own governments be provided in the UK, then the British Steel Corporation would be able to make an average 10 per cent reduction in prices, thus restoring its competitive power.

The required state subsidies would be as shown in Table 2.1.

Two variant forms of industrial commitment can thereby be identified, depending upon the stage of industrialisation. For Japan, it was indirect, extensive and coherent commitment reflecting a late and large-scale process of industrialisation, while for the UK, it was direct

Table 2.1 State subsidies required to restore BSC competitive power

Sector	*£million*
Coking coal	145
Rail freight	157
Labour subsidy	36
Research and development	22
Investment capital	132
Apprenticeship training	11
Oil for furnaces	2
Electricity	5
Total	510

Source: ISTC 1980: 52

commitment in the form of public ownership reflecting its later and more mature stage of industrialisation.

LABOUR MARKET[4]

The condition of the labour market is an important economic influence and varies according to the stage of industrialisation reached in any given economy. This section will review and compare employment in major divisions of economic activity and its relationship to labour in the steel industry.

The effect of an abundant and available workforce and its smooth shift from agriculture into the expanding manufacturing sector was more distinct in Japan than in Britain. The advantage accrued from it was crucial to the rapid and large-scale modernisation of the Japanese steel industry. In the period from 1960–73, employment in Japan increased by 7.98 million, while that of the UK rose by only 2.57 million. This illustrates the advantageous labour-supply conditions in Japan as a whole.

An examination of changes in employment patterns in the major divisions of economic activities shows that the workforce employed in agriculture, forestry and fishing, and mining and quarrying decreased sharply in Japan between 1960 and 1973, amounting to a loss of about 7.4 million people. Marked increases, on the other hand, were shown in manufacturing, trade, services and construction. The increase in employment among these latter four sectors during the same period amounted to 12.8 million. The swift and significant changes that occurred in employment patterns suggest a massive intra-divisional shift of the working population as a result of Japan's intense heavy industrialisation, taking the form of a great influx of workers into the manufacturing sector and a parallel increase of the working population in the trade and the service divisions. By contrast, UK employment in manufacturing and mining and quarrying divisions both fell sharply between 1960 and 1973, amounting to a decrease of about 1.1 million, and an employment growth of 1.98 million people is found in the trade and service divisions. Such changes suggest that the intra-divisional shift of the working population that occurred in the UK was characterised by a large fall-off (approximately 0.7 million people) in employment in manufacturing.

We therefore see that employment in the trade and service divisions increased markedly both in Japan and Britain. Finance, insurance and real estate also saw relatively large increases. Their total percentage share of employees in 1960 accounted for 34.2 per cent in Japan, and

38.6 per cent in the UK. By 1973 these figures had risen to 42.7 and 48.0 per cent respectively.

Two things can be deduced from these figures: first, the two economies show similar trends in the change of employment composition – a shift from primary to secondary, and then to tertiary industries. Second, in a comparison of the two economies during the period 1960–73, Japan industrialised rapidly and its workforce was mainly supplied from primary industries. Employment in manufacturing increased both in absolute and relative terms in Japan, while de-industrialisation in the UK proceeded both in absolute and relative terms, pushing her economy further towards tertiary industry.

By looking at changes among sub-sectors of manufacturing in the period 1965–73, we can see that in Japan employment increased in almost all industries in the manufacturing sector, while the UK suffered decline almost everywhere. In Japan a shift from light to heavy industries is more discernible than in Britain, although there is still a continuous flow of employment from light to heavy industries in the UK as well. A heavy concentration in the fabricated metal products and machinery industry suggests that profitability is relatively higher in this industry than in others, and this induces a constant inflow of capital and employment.

The above overview of the employment shift demonstrates that Japan achieved heavy industrialisation very quickly, thereby narrowing the gap with the UK. Similarities and differences in these employment trends indicate the existence of common forces in industrialisation itself, as well as national diversities in the timing, scale and velocity of industrialising processes.

Overall, the size of the steel industry (i.e. basic metal production) was relatively small in both countries, with less than 10 per cent of the workforce. Moreover, its relative position continued to decline. However, Japan accorded more importance to the steel industry than did the UK. General employment increased in Japan during this period, but continued to decline in Britain; detailed comparison shows that total employment between 1960–80 increased by 32,000 in Japan, but decreased by 205,000 in Britain.

The growth in Japanese employment does not include the numbers employed in sub-contracting firms, which expanded at a fast rate after 1960. For example, the sub-contracting rate for the whole iron and steel industry at the end of 1980 averaged 43.1 per cent, and that of the five major corporations was 47.6 per cent. Therefore total employment in the iron and steel industry that year, including employment in sub-contracting firms, amounted to 548,000 people; the five corporations themselves totalled 306,000 people. Compared to the number employed in 1960, this is an increase of 197,000 and 155,000 respectively. Thus, when sub-contracted employment is taken into account, the contrast in employment patterns between Japan and the UK becomes even more marked. As to trends, Figure 1.1 shows that employment numbers, after peaking in Japan in 1968 and the UK in 1961, began to fall.

As well as the change in numbers employed in the iron and steel industry, the proportion of employment in large-scale enterprises as a percentage of total employment has also changed. In Japan, employment growth in the five major steel corporations was larger than that for the industry overall, and the turning point from increase to decrease occurred later than in the rest of the industry. This suggests that the large steel companies have some difficulty in adjusting their workforce to respond to fluctuations in the corporate environment. The share of the five corporations within the total employed in the iron and steel industry continued to rise, to 54.0 per cent in 1980. In the UK, employment in the BSC actually decreased at a greater rate than in the steel industry as a whole, a process which has continued since. BSC's proportion of the industry workforce has also continued to fall.

Thus, the state of the labour market, as a factor of economic influence, has been important in determining the corporate environment of the steel industry. The Japanese steel industry was favourably positioned in terms of the labour market, allowing the easy formation of institutional arrangements, while the opposite situation prevailed in the UK due to the manufacturing industry as a whole being in decline.

STEEL IMPORTS/EXPORTS[5]

The assumption made in this section is that the general trend of import and export will exert economic influence and is relevant to decisions relating to capital investment. Competitive strength in exports reflects a stage of relative advance in production systems, while growth of imports reflects a stage of relative decline. The former indicates a situation of organic management compatibility, and the latter contrived compatibility. Once the direction of such competitive inertia has been determined, pressures from either imports or exports can become a stimulus for capital investment. When industrialisation advances to a level where steel demand stabilises and protective measures can be discarded, the stimulus for capital investment from within the domestic market will usually become smaller. But at the same time a competitive stimulus from overseas markets will come into play and companies will tend to continue their capital investment with a view to expanding exports in the foreign market as well as protecting the domestic market. It is at this point that the stage of absolute expansion gives way to a new stage – capital investment to compete and collaborate in foreign markets.

Let us first review the state of exports and imports of the two steel industries in order to compare their positions in the total trade of the two economies. The percentage of Japanese steel exports was relatively high in 1960–74, at an average of 13.4 per cent, and continued to rise until the 1970–4 period (15.2 per cent) when trade restrictions began to appear among importing countries. However, despite these, a high level of 14.6 per cent was maintained in 1975–9, firmly establishing steel's position as an export industry.[6] Steel had always been a minor element among total imports, but it fell even lower in the 1970s, showing the relative strength of the Japanese steel industry.

Yet in the UK, low export percentages tended to decrease further, while imports rose gradually, indicating a decline in the relative position of the steel industry as an export industry. Such an upward shift in the relative position of the Japanese steel industry in contrast to the UK suggests that once a situation of organic management compatibility has been established it increases its compatibility without much effort, while the opposite – that conditions of contrived compatibility must be consciously and strenuously maintained – is implied in Britain's case. The percentage of steel exports as a percentage of the total crude steel production in the UK shows a gradual rise from 18.3 per cent (1960–4) to 22.5 per cent (1970–4) and 25.5 per cent (1975–9), but this was due to a decrease of crude steel production keeping the existing level of steel exports in the same relative position.

The weak international competitiveness of the UK steel industry as an outcome of low labour productivity is discernible in the shift of the import–export ratio with the European Economic Community (EEC). The UK steel trade with the countries which were to form the EEC has expanded steadily since 1965, but the country became a net importer of steel products after the formation of the Common Market in 1967. In particular, for several years after 1973, when the UK joined the organisation, UK steel imports increased dramatically, as is reflected by the increase of the import–export ratio from 141.1 per cent (1971–3) to 331.1 per cent (1974–6). The underlying weakness of British competitiveness against the rest of the EEC is obvious, although the trade imbalance was much improved during the period 1977–9. As for importation of steel products from Japan, this was very small in 1960–4, at 33,500 tonnes, but gradually increased through the 1960s to 170,000 tonnes, and during the first half of the 1970s it jumped to over a million tonnes. This surge in imports stopped during the latter half of the 1970s, reflecting a restrictive tendency in trade relations between the UK and Japan.

The importance of exports for the Japanese steel industry, as shown above, implies that once export percentage exceeds a certain level, they become an important element for business operation, and competition in overseas markets among both foreign and domestic companies becomes a strong stimulus for capital investment. By 1979, the major steel companies had achieved within their total steel output export ratios of 32.9 per cent for Nippon Steel, 32.5 per cent for Nippon Kōkan, 38.9 per cent for Sumitomo Metal, 34.7 per cent for Kawasaki Steel and 24.5 per cent for Kobe Steel. Thus, for the large-scale steel enterprises of Japan, export has become both a cause and a result of plant and equipment modernisation and rationalisation.

British steel production differed from that of Japan in that its *weakening* international position encouraged the large-scale modernisation which was a major objective of nationalisation. The Labour government of 1964–70 issued a White Paper in 1965 in which it stated that one of the important tasks of nationalisation was to carry out plant and equipment modernisation and rationalisation. The modernisation of the UK steel industry at the stage of contrived compatibility was to be carried out by the deliberate application of nationalisation. Steel production came under public ownership in 1967, incorporating thirteen major private companies and one existing nationalised enterprise, Richard Thomas & Baldwins Ltd. The nationalised British Steel Corporation was one of the largest publicly-owned enterprises in the UK, with assets of about £1,400 million, an annual turnover of £1,000 million and 260,000 employees. It accounted for 94.2 per cent of liquid steel production and 66.1 per cent of steel products production. Nationalisation, however, implied a cessation of domestic competition, in particular the competition among the seven major enterprises whose liquid steel production had been over 1.5 million tonnes.

The Labour government which implemented the changeover was replaced by a Conservative government in 1970 and this change in government occurred before modernisation had properly got under way. Under the Conservatives, the Joint Steering Group was formed in 1971 by three organisations: the BSC, the Department of Trade and Industry, and the Treasury. This group published a White Paper on BSC's ten-year development programme in 1973, which included proposals for plant and equipment modernisation and rationalisation of the BSC. However, in 1974 the Labour Party was returned to power and under this government the programme was reviewed. In 1975 the Beswick Report was presented and proposed some changes to and postponement of the 1973 modernisation and rationalisation programmes. At last, in 1977, impelled by financial crisis at the BSC, full-scale rationalisation was set in motion.

From this experience it might be thought that the plant and equipment modernisation and rationalisation of the British steel industry, as represented by the BSC, began to take place some fifteen years later than that of the Japanese enterprises. The fact that plant and equipment modernisation and rationalisation of the BSC had been greatly delayed and was lacking in consistency due to political influences mainly caused by the changes of government was one of the important factors which interfered with BSC's chance to increase its competitiveness in both the domestic and overseas markets. For example, imports of finished steel to the UK show a sharp increase of 357.1 per cent from 1967–79; a slight increase may be observed in supplies from the private sector and a marked decrease in BSC's percentage share. This indicates that the competitive power of the BSC had been lost not only in the domestic market but also and to a greater extent in the international market. In addition, the fact that towards the end of 1977 the price of steel products had risen to almost equal that of the EEC countries and has since continued to rise is thought to have helped promote foreign steel imports. It was the surge in imports, reflecting the weakening of competitive strength in Britain, that finally acted as a pressure of competition to provoke the major rationalisation of the BSC which began belatedly in 1977.

CONCLUSION

This chapter has reviewed those factors in the corporate environment which have been relevant to corporate growth in the countries and period under study. It has been seen that macro-economic factors, such as general economic growth in the national economy and the prevailing state of the labour market, were major influences on corporate growth. To this may be added political influence as in the case of the modernisation and rationalisation of the British steel industry. Industrial policy, moreover, appears to have been induced by the stage of industrialisation reached in each country.

The conclusion that can be reached from this examination of the two steel industries is, therefore, that when a national economy is on an upward path and management can be characterised as having *organic* compatibility, as seen in Figure 1.3 (see p. 12) and exemplified by the case of Japan, macro-economic factors, along with certain other influences, stimulate the modernisation of industry. Britain's experience, however, suggests that when management is in a state of *contrived* compatibility, corporate environmental factors appear to be a negative influence on industrial renewal.

3 Industrial organisation

This chapter compares four dimensions of industrial organisation and considers them in terms of competitive incentives and pressures. We regard industrial organisation as a part of the corporate environment and assume from recent economic experience that Japanese industrial organisation seems to have been more conducive to successful competition than that of Britain.

OWNERSHIP AND CONTROL

Ownership and control is an institutional aspect of industrial organisation, and its relevance to managerial decision-making has a significant importance for capital investment. One characteristic of the ownership and control of large-scale iron and steel enterprises in Japan is that their major ownership is limited to a few financial institutions, with control in the hands of a top management class. A second characteristic, related to the first, is their relations with political elements, in particular with the ruling party and with the Ministry of International Trade and Industry (MITI), which has provided guidance, assistance and supervision.

As to the first point, the ownership of shares in large-scale steel enterprises is widely distributed, but the major shareholders are in most cases limited to a small number of banks and insurance companies whose holdings account for 20–30 per cent of all issued shares.[1] In contrast to the *Zaibatsu* (financial conglomerates) of the pre-war period, such an institutionalisation of shareholdings, together with the increase in size of the enterprises and sophistication and specialisation in the functions of top management, have led to the establishment of a professional managerial class.

The top management – in which we include the chairman, president, vice presidents, managing director and directors – are usually those who have been promoted from within the organisation; the ranking inside the top management is mainly based upon seniority in age and length of service.[2] They are the people who are believed to have combined a talent for leadership and logical thinking with respect for the importance of group harmony and a commitment to furthering the national economy. They are also people who have merited their positions through their contributions to the growth of the enterprise.

As to the second point, the Liberal–Democratic Party, in power since its inception in 1955, has maintained a strong relationship with the iron and steel industry by enacting legislation favourable to its growth, while at the same time receiving substantial financial contributions from it. For example, among the top thirty donors to the Liberal–Democratic Party, the major steel corporations and the Japan Iron and Steel Federation topped the list in 1976, accounting for a total of ¥468 million.

MITI, which has guided the reconstruction and development of the post-war economy, has given top priority to the iron and steel industry, thus providing various kinds of favourable treatment which were enacted through financial and taxation measures. For instance, MITI has maintained a long-term general policy aimed at making the iron and steel industry contribute to the development of the national economy through competition and co-operation in the markets. Competition among large-scale iron and steel enterprises has taken place under the administrative guidance of MITI. For example, in 1965 the ministry imposed sanctions against Sumitomo Metal Industries Ltd, reducing its allotted share of imported coal, for failing to abide by the agreement (Sumitomo Metal Industries 1967: 260). In addition to this, MITI played an important role, although not as vigorous as it used to, in promoting the amalgamation of the Yawata and Fuji iron and steel corporations, which took place in 1970 to form the present Nippon Steel Corporation.[3]

The identifying characteristic of the industrial organisation of the British steel industry is represented by the ownership and control of the British Steel Corporation. As a nationalised enterprise, it came under the direct supervision and control of the government. In the BSC, the members of the board, who were in charge of policy-making, and the members of the top management, who carried out the policies, were separate in terms of management organisation (see Chapter 10). That is, board members, including the chairman, were appointed by the minister and more than half of them were as a rule brought in from outside the BSC (as happened with the five chairmen appointed before March 1986, who were all from outside the iron and steel industry), while on the other hand the members of the executive top management were all promoted from within. Thus, professional managers from within the organisation have not been allowed to deal with decision-making in the way their Japanese counterparts were; in other words, their status within the organisation has been relatively low.

The 1967 nationalisation – in fact a renationalisation, following the post-war denationalisation of the industry by the Conservative government of 1953 – was carried out by the Labour Party. However, three subsequent changes of government, in 1970, 1974 and again in 1979, certainly had some influence upon the policies of the Corporation. On the whole this is reflected in the lack of consistency and continuity in policy and strategy. In particular, the Conservative government's economic policy since 1980 has gone so far as to aim at privatising parts of the BSC with a view to eventual overall privatisation of the corporation, as we will see in the final chapter. The BSC has thus experienced frequent change in the basic structure of ownership and control (BSC, Annual Report and Accounts 1983/84: 13).

There have been numerous examples of political gamesmanship with the BSC. Two examples of such direct political influence were the delay of the report by the Joint Steering Group in the formation of the Ten-Year Development Strategy (1973) and also the delay in executing the above strategy due to the Beswick Review. In addition might be mentioned various other direct and indirect influences in the areas of workforce rationalisation, organisational reform and price fixing (Silberston 1978: 140–51; 1982: 99–102).

Thus the difference in the structure of ownership and control, and its subsequent effect on the attitude of top management, have influenced decision-making towards capital investment. We can assume that the Japanese style of ownership and control was a reflection of positive compatibility, which allowed greater freedom to management, whose preference was growth; while that of Britain was a reflection of contrived compatibility, which placed management under strict control of a board whose preference was profitability.

COMPETITIVE STRUCTURE

Competitive structure is an important dimension of industrial organisation and it determines the degree of competitive pressure. In Japan, under the administrative guidance of the MITI, the large-scale steel enterprises, run by professional managing groups as major elements of larger financial groups, have maintained oligopolistic competition. The competition is characterised by that of prices skilfully guided by the MITI, product quality, innovative products, shortened and exact delivery times, and by the establishment of close ties with customers in the same financial group. As a result of strong competition in the above areas, the major steel companies were forced to modernise plants and equipment, to expand production, and to strengthen international competitiveness.[4]

First, the nature of this competition is specifically reflected in the process of concentration of production of crude steel and ordinary hot-rolled products in the hands of the five major corporations. In the decade from 1960–70 the production share of these companies in these products increased from 69.6 to 78.7 per cent in steel and 67.7 to 77.8 per cent in rolled products, thus allowing even stronger control to be exerted by the major five corporations over the whole steel industry. Although a slight decrease is recorded in the corporations' percentage share of the production of crude steel and ordinary hot-rolled products in 1979, this was due to a rise in the share of electric furnace enterprises.[5] Second, although Nippon Steel's percentage fell, and those of Sumitomo Metal and Kawasaki Steel rose in all categories of iron, crude steel and ordinary hot-rolled steel products, Nippon Steel's share still remained far larger than those of the other enterprises. In consequence, as Nippon Kōkan, Sumitomo Metal and Kawasaki Steel came to stand in similar relative positions, the structure of competition has been changed to make it easier for the Nippon Steel Corporation to exercise its leadership in setting prices and encouraging non-price competition (Okishio and Ishida 1981: 146–50).

The present structure of competition was given an impetus when the then major open-hearth furnace enterprises, Kawasaki Steel, Sumitomo Metal and Kobe Steel, adopted an integrated iron and steel production system in the early half of the 1950s.[6] Kawasaki Steel specialised in iron and steel production when the iron production department was separated from Kawasaki Heavy Industries in 1950. Sumitomo Metal also specialised in iron and steel production when it detached its Copper and Aluminium Rolling Department in 1959, Aircraft Department in 1961 and Magnetic Steel Department in 1963. Kawasaki Steel and Sumitomo Metal, which specialised only in iron and steel production, began to build new integrated works ahead of the other enterprises.

Taking advantage of increasing demand both inside and outside Japan, Kawasaki Steel and Sumitomo Metal obtained and have increased their market share of steel products which until then had been monopolised by the existing integrated iron and steel enterprises (Yawata Steel, Fuji Steel, Nippon Kōkan). MITI's administrative guidance promoted competition to the extent of preventing bankruptcies and allowing for equal opportunity in providing credit loans from the

Bank of Japan to each financial group. Under conditions of high economic growth, Kawasaki Steel and Sumitomo Metal, which concentrated their investment on steel production in new integrated works, were in a more advantageous position to expand their market share than Nippon Steel, which owned older works, and Nippon Kōkan and Kobe Steel, which owned other business departments such as engineering and shipbuilding. But competition in the period of low economic growth that commenced in 1974 has created new areas of competition, maintaining the price co-ordination led by Nippon Steel. This is a reflection of managerial efforts at the stage of contrived compatibility. The new areas where competition is mainly concentrated are in measures to reduce costs, save energy, rationalise personnel and strengthen small group activities; the development of new areas of business (engineering and chemical departments, and overseas activities); and the promotion of research and development. These new areas are important when considering the competitiveness of these enterprises.

Competition among the major steel companies stimulated modernisation of plant and equipment, specifically mass production systems, by establishing modern integrated works. As a result, the competitive strength of the Japanese steel industry (in terms of price, quality, range of products and delivery) increased markedly, and as shown in the market share of finished steel to Japanese consumers and merchants, the domestic demand was met almost 100 per cent by Japanese companies all through the period 1960–80.

In the British iron and steel industry, competition among the seven major enterprises responsible for more than 1.5 million tons of crude steel disappeared on nationalisation in 1967. The competitive environment in the iron and steel industry shifted to competition between the BSC and remaining private enterprises in the UK and competition between the BSC and overseas enterprises, in particular those in the European Community.

In the competition between the BSC and private enterprises in the UK, the market share of private enterprises increased from 24.4 per cent in 1967–9 to 27.0 per cent in 1979. At the time of nationalisation there were about 130 independent private enterprises, of which thirty produced around 14 per cent of total crude steel output. Of the crude steel they produced, one quarter was alloy steel and the remainder was non-alloy steel (Cockerill 1980: 135). Three reasons may be advanced for the added competitive power of the independent private enterprises: the conditions for securing scrap had improved; the private producers succeeded in increasing their productivity; and the reliability of BSC's delivery had decreased (Cockerill 1980: 142).

In the area of competition with foreign enterprises, the domestic market was protected before entry to the EEC by government intervention in pricing and taxes on imported steel products. Consequently prices for steel products were kept lower than those in the international market. Thus for a time the BSC enjoyed conditions favourable to the sales of her products. But entry to the European Community meant a gradual elimination of such favourable conditions, so that prices by the end of 1977 were at parity with those in the Community, and on some products were even higher than EEC levels (Cockerill 1980: 143). Such a situation is shown clearly by trends of steel imports and exports between the UK and the EEC; steel imports from EEC countries have increased drastically since the UK entered the Community, while exports to them have remained by and large constant, thus showing a relative decline in the international competitiveness of the BSC.[7]

During the long recession of the 1970s, competitive pressure from so-called third world countries increased, creating a further obstacle to regulating steel prices in the Community. Towards the end of the 1970s steel crisis measures were introduced, as will be explained in detail in the next section. The aim of the crisis measures was to create co-operation and co-ordination among the enterprises in the Community and at the same time protect them against outsiders. However, despite such a co-operative and co-ordinated competitive environment, the position of the BSC has still declined relatively, in terms of increasing percentages of imports and decreasing market share.[8]

Thus, in comparison, we can assume that the oligopolistic structure of Japan was more suitable than the nationalised structure of Britain for creating stronger competitive pressures for technological innovation. The nature of the competitive structure which Japan enjoyed seems to have become an agency for increasing positive compatibility of management.

RESTRICTIVE AGREEMENTS

The extent and degree of government commitment to industrial organisation differs depending upon whether a stage of positive or contrived compatibility is operative, and it affects the managerial strategy for capital investment. MITI-managed competition has been a characteristic element of Japanese industrial organisation. It has been useful in producing overall and long-term benefits for the major steel companies. MITI has always taken the initiative in encouraging co-operation in response to prevailing economic situations. In this section, the 'open sales system' which aimed to stabilise prices in the 1960s, the 'voluntary adjustment' of plant and equipment modernisation and the 'price leadership' since the formation of the Nippon Steel Corporation will be examined as some of the restrictive agreements employed.

The Open Sales System was introduced for the first time in 1958 under the administrative guidance of the MITI and was a countermeasure against recession, aiming at price stabilisation and regulation of production (the First Open Sales System). It was subsequently modified in May 1960 to a countermeasure against boom (the Second Open Sales System) and in July of the same year it was again modified to function as a regular price stabilisation measure (the Third Open Sales System). The aim of the Third Open Sales System was as follows:

> In order to secure stable growth for our economy, the announcement of sales prices for major steel products, the adjustment of output and volume of sales, and implementation of simultaneous sales shall be made. By preventing marked changes in demand and supply as well as in steel product prices, the stabilisation of prices at a lower level shall be achieved.
>
> (Tekkō Renmei 1969: 788)

This Open Sales System, although it became nominal due to the sharp decline in the price of sheet steel which occurred in the middle of the 1960s, can nevertheless be said to have played an important role in bringing about a co-ordinated reduction of crude steel output at times of recession (Tekkō Renmei 1969: 202). The eventual collapse of the Open Sales System indicated increased competition among the major steel companies, and this increased competition seems to have promoted the amalgamation of Yawata and Fuji corporations which took place in 1970.

The adjustment of plant and equipment was inevitable to secure any stabilisation in prices. The steel enterprises met at the end of 1959 to encourage the voluntary adjustment of plant and equipment investment (Tekkō Renmei 1969: 56–7). However, it became apparent that strong differences of opinion existed on the criteria for establishing voluntary adjustment between the already existing large-scale integrated steel enterprises and the newcomers. The former enterprises, such as Yawata and Fuji steel corporations, maintained that any permission to upgrade plant and equipment should be based upon the size of existing market share, while the latter group, such as Sumitomo Metal Industries Ltd and Kawasaki Steel Corporation, insisted that it should be based upon the demand forecast for each product and also upon the financial condition of each company (Ichikawa 1969: 211).

A decision concerning voluntary adjustment for capital equipment still had not been reached by 1962 and it was finally left to MITI to co-ordinate resolution. Yet still the result of this voluntary adjustment was not effective. In the face of the steel recession of 1965 the necessity for plant and equipment adjustment was again felt strongly; in July 1965 an agreement for voluntary adjustment was finally reached, but no regulation of investment for blast furnaces, LD converters and ingot-making equipment was achieved, thus indicating the intensity of competition among the major enterprises. In 1967, 1968 and 1969 voluntary adjustments for blast furnaces were agreed upon, but in practice the companies competed with one another in building new blast furnaces.

With the formation of the Nippon Steel Corporation in 1970 the competition of the 1960s changed to a more co-operative attitude, as shown by the firm leadership of the new corporation. For example, a decline in prices has not occurred in the post-1970 recessions as it had before, and moreover prices have been maintained or even raised, while at the same time a co-ordinated reduction in production has been achieved. In the matter of raising prices the Nippon Steel Corporation has always taken the initiative, followed by the others. In addition, administrative guidelines from MITI, such as those relating to the increase and decrease of crude steel output and steel products production have been adhered to and thus have proven effective among large-scale steel enterprises.

In Britain, where the BSC was left as the only large-scale iron and steel enterprise subsequent to nationalisation, such restrictive measures among large-scale enterprises as existed in Japan disappeared. Instead we find government control of prices in the period prior to entry to the EEC, and subsequently co-ordination of BSC policy under crisis measures undertaken for the steel industry throughout the European Community.

A major objective of the government's introduction of price fixing was to use it as a means for combating inflation. BSC prices were always held lower than European market levels. The following are some examples to show the government's refusal to grant the BSC's request for price rises:

1969: Price rises requested in December 1968 not granted in full until November 1969.
1970: Requested price increase delayed three months.
1971: Average price rise of 14 per cent requested but the Government advised BSC to halve it.
1971–2: BSC allowed only a 4.6 per cent increase, and this was deferred until April 1972.
1973: Start of removal of price controls.

(ISTC 1980: 69)

The price levels for steel products in early 1973 were estimated to be about 15 per cent lower than those for the ECSC (Federal Trade Commission 1977: 443). Such low price levels, which might have contributed to the decline of steel imports and also of the steel consuming industries in the UK, became a cause of financial loss to the BSC. If the price level of the ECSC had been maintained in the UK during 1967–75, it is estimated that £783 million would have been received in further revenue (ISTC 1980: 60). Upon entry into the Community, the ECSC basing-point price system was introduced in April 1973. Since then, prices have been on the increase and, as explained above, by the end of 1977 price levels in the UK were almost the same as those in the Community. The prices of some products which were competitive were even higher than those in the Community at that time.

The steel crisis in the European Community became more serious in the latter half of the 1970s. The crisis was revealed in the reduction of crude steel production, the fall in prices, the decrease in employment and the considerable financial losses. In order to cope with this situation, Eurofer (a new organisation of the European Community steel producers) was established in December 1976 to carry out crisis measures planned by the European Commission. The first of these measures was the Simonet Plan (January 1977) which involved the voluntary limitation of deliveries within the Common Market of certain products (Nippon Steel 1978: 30–40; BSC, Annual Report and Accounts 1976/77: 11). The Davignon Plan was implemented in May of the same year in order to re-inforce the Simonet Plan. The following measures were included in it (BSC 1976/77):

1 the institution of guideline prices;
2 an automatic 'surveillance' licensing system for imports;
3 community finance for running down old capacity and for modernisation.

As the Davignon Plan did not bring about the desired results, the European Commission introduced an even more comprehensive set of counter-crisis measures prepared by Vicomte Davignon in January of 1978. The following measures were set forth:

1 the extension of mandatory minimum prices to merchant bars and hot-rolled wide coil;
2 a temporary system of import reference prices (applicable to producers and stockholders at a margin below domestic price levels), below which imports would be liable to a provisional anti-dumping duty;

3 an increase of about 5 per cent in the guidance prices.

(BSC 1977/78: 10–11)

As various voluntary counter-crisis measures based upon co-ordination and co-operation in the latter half of the 1970s had not brought about satisfactory results, in 1980 the 'manifest crisis' policy, based upon Article 58 of the Paris Treaty, was issued and in November of the same year a regime of compulsory production quotas was introduced (BSC 1980/81: 13; Commission of the European Communities 1982).

The above-mentioned series of counter-crisis measures by the European Commission was designed to bring about the temporary stabilisation of prices and, more importantly, to restructure the steel industry in the Community. Therefore the rationalisation of the European steel industry in this period, including the BSC, was remarkable. Co-ordination and co-operation in fulfilling production quotas in the Community is at the same time an aspect of competition among the Community iron and steel enterprises. The BSC promoted its own rationalisation by taking advantage of the measures implemented by the European Commission to reduce production.

Restrictive agreements, as part of a general industrial organisation, were thus used in Japan as a tool for co-ordination among major steel companies in the interests of competitive growth, while in Britain such action was mostly taken in the context of the European market and its co-ordination was for the sake of contrived compatibility of management.

MERGERS/CONCENTRATION

Mergers and concentration reflect a process of changing industrial organisation, and are consequential to intensity in competition. They serve to effect a change in the nature of industrial organisation towards an oligopoly and create larger business organisations which require sophisticated management systems.

The evolution of mergers or concentration in the Japanese iron and steel industry was remarkable throughout the 1960s and reached a peak with the amalgamation of the Yawata and Fuji Corporations in March 1970. The post-war period saw the following business aggrandisement activities by the large-scale steel enterprises:

1 acquisition of Ogura Seiko by Sumitomo Metal Industries Ltd (1953);
2 acquisition of Amagasaki Seitetsu by Kobe Steel Ltd (1965);
3 acquisition of Tōkai Seitetsu by Fuji Iron and Steel Corporation (1967);

4 acquisition of Yawata Kōkan by Yawata Iron and Steel Corporation (1968);
5 merger of Yawata and Fuji Iron and Steel Corporations (1970);
6 acquisition of Fuji Sanki Kōkan by Nippon Steel Corporation (1971).

The above acquisitions and mergers were in each case carried out to compensate for weak competitiveness in certain product areas, and were thus aimed at increasing comprehensive capacity as an integrated steel enterprise. While these mergers were initially the result of intense competition among the large-scale steel enterprises, they have in their turn generated pressures for further competition.

The competition entered a new phase in 1970 with the amalgamation of the Fuji and Yawata Corporations to form the Nippon Steel Corporation. This amalgamation will be examined from two angles: one concerning the aim of the amalgamation and the other the impact that it has had upon the whole structure of the iron and steel industry.

Merger was expected to bring about the following results (Nippon Steel 1981: 120–3):

- increased efficiency in plant and equipment investment;
- increased efficiency in R & D investment;
- increased efficiency in the purchase, transport and use of materials;
- increased efficiency in the distribution of production;
- increased rationalisation in product shipping;
- improved service and full sales;
- rationalisation and computerisation in management departments;
- increased capacity for capital procurement;
- increased overseas business activities;
- increased overall capacity of the enterprise.

It must be asked if these objectives were actually achieved. Due to the world-wide recession since 1973, domestic demand has not increased, so it would be wrong to say that the expected results have been substantially realised. The lower level of equity capital, insufficient profits and the decrease in the market share of iron, steel and steel products support this view (Nippon Steel 1981: 381). But in the areas of overseas technical assistance and engineering activities, the goals of amalgamation may be said to have been thoroughly achieved (Nippon Steel 1981: 383–4).

As regards the second point, the formation of the Nippon Steel Corporation produced a previously unseen level of co-operation among the major steel companies. This is shown, for example, in the restraint in price competition created by the price leadership of the Nippon Steel

Corporation. Indeed, they succeeded in raising the price for four years in succession from 1974 to 1977 (Tekkō Renmei 1981: 195–9).

Another phenomenon which was observed in connection with the change in the nature of competition was the progress in reorganisation of small- and medium-sized enterprises involved with electric furnaces and those in special steel production. This reorganisation was achieved by grouping these enterprises together, with the major steel companies providing capital, technology and personnel assistance. For example, the Nippon Steel Corporation helped to promote mergers and amalgamations among small- and medium-sized enterprises eight times in the 1970s and it has also increased its own *Keiretsu Kigyō* (affiliated enterprises), thus expanding its own enterprise group (Nippon Steel 1981: 548–53).

In the UK, the share of production of those fourteen companies which were later nationalised for the period of July 1963 to June 1964 accounted for 96 per cent of iron, 94.5 per cent of ordinary crude steel, 63 per cent of alloy steel, 90 per cent of plate, 97 per cent of heavy section and bars, 98 per cent of sheet and 100 per cent of tinplate, thus showing a very high concentration in these items. The percentage share is lower in other items, such as tubes, pipes and fittings (71 per cent), wire rods (55.5 per cent), forging, tyres, wheels and axles (42 per cent), light sections and bars (50.5 per cent), cold-rolled strips (34.5 per cent) and bright steel bars (10.5 per cent) (Cmnd 2651, Ministry of Power 1965).

The nationalisation was intended first, to stimulate general economic development, second, to provide finance for modernisation programmes, and third, to increase competitive power to promote exports. These aims of nationalisation suggest that they had not been adequately met through market mechanisms prior to nationalisation. Thus the goal of nationalisation was to achieve the revitalisation and development of the British steel industry by means of public ownership.

Again we must look at the figures to ask if these objectives have been achieved. Although the total amount of plant and equipment investment in the period 1970–80 was £4,177 million, crude steel capacity actually fell, crude steel output halved and employment dropped by 130,000 people (see Table 8.1, p. 158). Since the position of the BSC more or less was the British iron and steel industry, the above index of general performance applies also to the BSC.

In addition, BSC's share of the domestic market in the period from 1967–79 decreased by 30 per cent while the private sector share increased by 10.7 per cent. Imports increased markedly in the same period by as much as 257.1 per cent. In terms of relative market share in 1979, BSC accounted for 48.0 per cent, the private sector

27.0 per cent, and imports 25.0 per cent, thus showing a marked relative decline of the BSC. The above facts indicate that the aims and roles expected of the BSC have not been satisfactorily achieved.

Thus, the competitive pressures in Japan changed the nature of industrial organisation, while it was political decision in Britain which changed the nature of industrial organisation. In both cases they resulted in the formation of huge enterprises, which in time led to a relative decline in their market shares.

CONCLUSION

The changing nature of industrial organisation in the Japanese and British steel industries has been reviewed from four perspectives. It has been seen that changes in the two countries resulted from differing processes, namely competition in Japan and nationalisation in Britain, but both tended in the end towards an increase in scale and an oligopolistic character which weakened competitive pressures in the domestic market. These variant processes of adjustment also reflect management processes, again illustrating organic compatibility, a compatibility in the rising trend of industry in Japan and contrived compatibility, a compatibility in the declining trend in the UK.

4 Institutional practices

Within any business organisation, the institutional practices of today exert a determinant effect upon the decision-making of tomorrow. Institutional practices exist in a constant process of continuity and change, and hence exist as both cause and result simultaneously. Regarding given institutional practices as constituting an internal corporate environment conditioning future managerial strategies, this chapter investigates and compares four institutional factors and considers their relevance to capital investment, and the introduction of new technology.

EMPLOYMENT

The nature of employment practice is a reflection of the relationship between an individual and business organisation, and differs depending upon the stage of industrialisation.[1] In this context, Japanese employment contracts are more collective and all-embracing, while UK contracts are more individualistic and job-specific. In Japan it is usual to recruit new employees collectively without specifying job or pay, which means that newly recruited employees begin their working life as beings who can be used and developed as human capital at the company's discretion,[2] while British workers are employed as individuals whose job and pay are set out in their contracts.

A major characteristic of the employment practice in Japanese large-scale steel enterprises has been the principle of employing young graduates from various educational careers on a long-term basis, without specifying particular jobs. The implication is that they join and belong to the company and continue until retirement to work within and for it under a broad category of blue- or white-collar work (Dore 1973).

Companies recruit various employees by academic career every year in April. The potential recruits are high school graduates (18 years old), higher professional school and junior college graduates (20 years old), university graduates (22 years old) and postgraduate degree students (24 years old). In Japan, as the level of wages and salaries for the large-scale steel companies were relatively high during the 1960s, they succeeded in recruiting the pick of the graduates from schools and universities.

The young inexperienced graduates are given an introductory education and training depending upon their school career and then are allocated to an appropriate workshop or office to undertake a job assigned there. A majority of male high school graduates will become manual workers, female high school graduates and junior college graduates will undertake general office work, male higher professional school graduates and university graduates will take up office or technical positions (depending upon what they have studied in school and university), and those who have finished a postgraduate course go into R & D work. The education and training upon entry is all done within the enterprise and is systematic and group-oriented. Although the length and content of induction education and training varies according to school career, the major aim is to establish an attitude of commitment and through this it is expected that a collective and loyal commitment to the enterprise may be formed (Tekkō Renmei 1962: 3).

Promotion is implemented in a systematic and long-term fashion on the assumption of life time employment, which implies employment until compulsory retirement age. The prerogative held by management for personnel evaluation and promotion is extremely strong and not subject to any union regulation. As recruitment is regular and limited to new graduates, age and length of service correspond for the most part, but it does not mean that people of the same age and the same length of service are always promoted to similar managerial positions. For manual workers to be promoted to the position of group leader and foreman, and for clerical and technical employees to the position of sub-section manager, section manager or department manager, they have to pass a strict personnel evaluation by the management. Those who satisfy the requirements of age and length of service and yet are not promoted to a corresponding managerial position are guaranteed a certain level of increase in wages and salary according to 'qualification' (level in grading) by personnel evaluation. This 'qualification' system is an important device to solve the problem of the gap between the number of managerial positions and the employees who qualify for them on the basis of seniority.

Education and training are provided systematically even after the induction education and training as an important means of personnel management. This aims at providing employees with the necessary professional knowledge of each particular field and helps them understand the objectives, policy and strategy of the enterprise, thus preparing them to accept any kind of redeployment, and also giving management unilateral control over the internal labour market. For example, Yawata Steel Corporation effected large-scale redeployment to create new, modern, integrated steel works at distant locations. These involved shifting 1,269 people from Yawata Works to Hikari Works (1954–69), 2,936 people to Sakai Works (1959–69) and 3,163 people to Kimitsu Works (1964–9) (Nippon Steel 1980: 455).

The nature of the employment contract in Japan corresponds to the type of in-enterprise education and training and also to methods of promotion and assessment. It further corresponds with labour management practices, such as the easy transfer of employees from one works to the other. Such an employment contract has arisen mainly because of the stage and process of Japanese industrialisation, which can be characterised as a late and rapid process which determined the structure of the labour market and the type of managerial strategy for the appropriate marshalling of labour resources.

A major employment characteristic of the BSC, in contrast to the Japanese enterprises, was that employees had an explicit bilateral employment contract, with the Corporation specifying a particular occupation and a corresponding wage. In effect, the conclusion of an employment contract implies the selling of one's own labour power to perform a particular job for a particular level of wage or salary. This job-specific contract philosophy works against the 'family' idea of an enterprise as a place to develop employees, as distinct from a place to earn money. Employee loyalty in this latter case, if desired by management, would need to be created by some more immediately realistic means. Job- and pay-specific employment inevitably obliges management to fill a vacancy as soon as it occurs; this requires employment from outside or, if the post is to be filled internally, readjustment of a wage agreement. If employment was as regular a practice as it is in Japan, vacancies arising would have been left unoccupied until the next recruiting time.

In addition, in Britain certain positions are subject to a 'needed personnel' agreement with the trade union, by which a vacancy must be filled as soon as it arises.[3] The system of filling vacancies whenever they occur removes the need to limit new employees to recent graduates and/or inexperienced persons and rather encourages the recruitment of experienced personnel. To be employed as a skilled worker, one must have an appropriate qualification – usually a four- or five-year apprenticeship – but ordinary and ancillary workers require no examination or official approval apart from physical suitability (Tekkō Renmei 1974b: 57).

Education and training within the enterprise is therefore limited only to the required skill and knowledge necessary for the employee to carry out the job and as a means for promotion, thus differing from the Japanese counterpart which puts emphasis upon collective and systematic education and training. No attempt is made to inculcate values fostering an individual's commitment to the enterprise as a whole; much of the education and training of the worker is performed by extra-industrial public institutions.

Education and training are a condition for promotion for managerial employees, but do not offer any incentive for manual workers in that sense. Operatives, owing to the seniority system controlled by the union, have the order of job promotion ready-determined, and other manual workers, through traditional practice, are broadly classified into skilled, semi-skilled or unskilled. Recruitment of managerial staff is done by public advertisement, describing clearly the qualifications required, level of salary offered and job character. Thus, promotion and redeployment are not executed by the unilateral prerogative authority of management. Promotion and movement within senior management are realised through individual negotiation with the personnel director as to the level of salary, job character and conditions. Here we can observe a manifestation of the individualistic philosophy so characteristic of the traditional middle class ethic (Hazama 1974).

Thus, in comparison, we can see that the employment contracts of Japan and Britain differ in nature and the distinction is relevant to the style of labour management in the two steel industries. In Britain, a narrow specification of job and pay, together with a relatively strong trade union influence, seems to have limited the scope of managerial strategy in the use of labour resources, while a broad relation between job and pay and a long-term perspective of employment in Japan seem to have allowed considerable managerial freedom in the flexible use of their workforce. This situation has been an important constituent of the organic compatibility of management in Japan.

STRUCTURE OF THE WORKFORCE

The delayed, rapid and government-supported industrialisation in post-war Japan created a unique feature, a distinct and relatively large-scale, dual-structured workforce. The structure recognised in the large steel companies was that the number of temporary employees began to decrease around 1960, while in-works sub-contracting began to increase, thus creating a dual workforce structure of *honkō*, the employees of the iron and steel companies, and *shitaukekō*, the employees of sub-contracting firms (Nippon Steel 1980: 324–36).[4]

Even before 1960, in-works sub-contracting existed in the areas of in-works transportation and raw material processing, but since 1960 it has expanded into areas of work incidental to iron and steel making and that of processing and manufacturing, and even into sub-contracting a whole process (Nippon Steel 1980: 324–36). The percentage of sub-contracting is larger for new large-scale integrated works than for an old works. The rate of sub-contracting has since further increased, but in 1960 was 27.3 per cent on average for Yawata Steel and 24.1 per cent on average for Fuji Steel (Nippon Steel 1981b: 326).

In-works sub-contracting firms can be broadly classified, depending upon the type of work they supply, into regular work sub-contracting firms and engineering sub-contracting firms. In the Kimitsu works of the Nippon Steel Corporation there are forty-two of the former and seventeen of the latter (see Table 5.7, p. 96). An old large-scale works such as the Yawata works of the Nippon Steel Corporation had a lower sub-contracting rate of 44.4 per cent in 1976, but the actual number of sub-contracting firms was much larger, at 128 regular work sub-contracting firms and sixteen engineering sub-contracting firms (see Table 6.3, p. 103).

The comparison of average age, length of service, wages and working hours between *honkō* and *shitaukekō* is revealing. In the case of the Hirohata works of the Nippon Steel Corporation in 1975, the average age of *honkō* was 36.4 years, while that of *shitaukekō* was 41.2 years; the average length of service for *honkō* was 15.2 years, while that of *shitaukekō* was 9 years; the average level of wages, including overtime, for *shitaukekō* was 85.3 per cent of the *honkō* pay, while their seasonal bonus for summer and winter was 81.4 per cent of the *honkō* figure; the retirement lump sum payment for *shitaukekō* with twenty-five years length of service was 61.5 per cent of the *honkō*, and the average monthly working hours of *shitaukekō* stood at 215.7 hours compared to 166 hours for *honkō*, which amounts to 29.9 per cent less (Imada 1981: 231–6). In addition, more *shitaukekō* died in labour accidents than *honkō*.

Such realities of the in-works sub-contracting system show that various functions which could primarily have been executed by the steel companies are being successfully fulfilled by the use of sub-contracted employment as an intra-enterprise division of labour. The large steel companies not only save labour costs by the use of sub-contracting, but also gain various other advantages:

1 to shift the role of employment adjustment necessitated by economic cycles to the sub-contracting firms;
2 to shift the complex industrial problems created in areas where physical and monotonous labour are abundant, even with a higher rate of labour accidents, to sub-contracting firms, and consequently to stabilise their own industrial relations;
3 to provide their own employees with a sense of superiority, self-consciousness of status and enterprise orientation by dint of their better labour conditions than those held by the employees of sub-contracting firms. This contributes to raising the sense of loyalty within the enterprise and fosters co-operative industrial relations.

(Ishida 1967: 104)

The dual nature of workforces in the Japanese steel industry has thus been an important consequence of its rapid development since 1960, becoming a ubiquitous part of the corporate environment.[5]

A major characteristic of the workforce structure of the British Steel Corporation is that the workforce is classified into four occupational groups: managerial, technical and clerical employees; operatives; maintenance and craftsmen; and ancillary employees. The percentage of each category at the Scunthorpe works in 1972 were 24.3 per cent, 34.5 per cent, 26.4 per cent and 14.8 per cent respectively (see Chapter 8, pp. 165–75).[6]

The group of managerial, technical and clerical employees includes directors and managers, management in special functions, scientists and technologists, functional specialists, supervisory employees and ordinary clerical employees. The group of operatives includes all the operatives at works and workshops; they are given a job ranking order and are assigned to jobs on a seniority basis. Each job has a different wage rate. At the bottom of the order there are junior operatives whose wages are very low. The maintenance and craftsmen group includes skilled and semi-skilled employees and other workers and apprentices; skilled workers account for about 50 per cent of all employees of this group. There are about twenty skilled occupations and the rate of wages is determined for each occupation. The group of ancillary employees includes crane drivers and some other occupations.

Although there are these four broad occupational groups, we can recognise some further differences in employment conditions between white-collar employees (managerial, technical and clerical) and blue-collar employees (the other three, manual groups). These occur in the amount and form of payment, working hours, annual holidays, sick pay and pension rights. In addition, there have existed distinctions in the use of toilets, canteens, language, clothes, hobbies and values, which are related to social differentiation associated with traditional class relationships.[7] Among white-collar employees there are also some employee groups which reflect their occupations and positions and this is also true of blue-collar employees, where such groups form the basis of various trade unions. As a result, plant and equipment modernisation and rationalisation which lead to increase or decrease of occupations or changes in the content of jobs tend to become a source of demarcation problems among trade unions, thus making industrial relations all the more complex.

The wage differentials among occupational groups in 1972 revealed that, compared to the average level of salaries for supervisors, technicians and clerical employees, that of process operatives was 96.6 per cent, ancillary employees 95.0 per cent and maintenance and craftsmen 89.6 per cent; however, the average wage of the skilled employees, who are included under maintenance and craftsmen, is itself higher than that of any other group.[8] Therefore, although there exist average wage differentials among groups, that degree of difference is actually rather small. The existence of differentials in wage and salary among individual occupations and jobs seems to be an important British characteristic.

As to age composition and length of service, managerial, technical and clerical employees are higher in age composition and longer in length of service than blue-collar employees, so that white-collar employees show a trend of long-term employment similar to that of Japan. Among blue-collar employees, the age composition of the three occupational groups is similarly high, but when this is examined in detail it can be seen that length of service is relatively high for process operatives who are under a union-constrained seniority system, and for maintenance and craftsmen organised in crafts unions and classified into traditional categories of skilled, semi-skilled and unskilled employees.[9] The length of service in ancillary employment, where jobs are mostly carried out by unskilled and female labour, is extremely short.

The two contrasting workforce structures, once established, seem to have functioned in different ways. The Japanese structure has operated favourably for the managerial strategies aimed at corporate growth in a general climate of organic compatibility, while the British structure has performed unfavourably for managerial strategies aimed at simultaneous rationalisation and modernisation.

WAGE, SALARY AND PROMOTION

Wage, salary and promotion are a reflection of corporate social relations, but at the same time function as a tool for integrating factors of management such as work practices, employment practices, working conditions, industrial relations, technology and corporate growth, which have already been explained in Chapter 1 under the discussion of system compatibility.

There are four major features of the salary and promotion systems of the major Japanese steel companies. First, the general salary level in the large steel enterprises has been higher than that of other industries. In particular, in the period 1965–70 salaries rose to a point in 1970 where in all age ranges those clerical workers joining straight from university or high school, and even manual workers joining from high school (except the age range 35–39) enjoyed higher salaries than their counterparts in any other industry (Ishida 1981: 51–4).

Second, the level of salaries in the large steel companies is higher than that in small- and medium-scale steel enterprises. For example, the average salary for those in small and medium enterprises was 85.1 per cent of that in the five major companies in overall terms, 85.0 per cent of the initial salary for university graduates and 88.0 per cent for high school graduate manual employees. Other kinds of earnings, such as seasonal bonuses and the retirement lump sum payments, are also smaller for those in the small and medium enterprises (Tekkō Rōren, *Rōdō Handbook* 1966).

Third, within the large steel companies there exist salary differentials based upon school career. For example, in 1965 initial salaries, compared to those of university graduate technical and clerical employees, were 63.2 per cent for high school graduate technical and clerical employees (male and female) and 68.3 per cent for high school graduate manual employees (Tekkō Rōren, *Rōdō Handbook* 1966). This differential in initial salary continues to widen in accordance with length of service, reaching at retirement (50–59 years old) 59.9 per cent for high school graduate technical and clerical employees (male), 32.7 per cent for female and 37.7 per cent for high school graduate manual employees. Since 1965 this gap has reduced, but the differential was still 67.5 per cent, 40.0 per cent and 50.1 per cent respectively in 1978 (Ishida 1981: 60).

Fourth, the percentage of salary determined by age and length of service has always been around 50 per cent and the increase in earnings due to promotion to the higher rank of managerial positions is paid in the form of managerial rank allowance. The rise of salary determined by age in 1965, taking the 20–24 years age group as 100 per cent, was as follows in the retirement age group (50–59 years old): university graduates, technical and clerical 472.1 per cent, high school graduates, technical and clerical (male) 325.2 per cent, high school graduates, technical and clerical (female) 213.0 per cent, high school graduates, manual employees (male) 199.2 per cent. This age–seniority element in the wage/salary system has tended to fall since then but was still maintained in 1978, accounting for 359.2 per cent, 274.4 per cent, 171.9 per cent and 185.1 per cent respectively (Ishida 1981: 55).

The whole salary package consists of four major elements: basic rate, job and capacity rate, efficiency rate and various allowances. The basic rate is determined by seniority elements such as school career, age and length of service. The job and capacity rate was introduced on the advice and guidance of the United States in 1946 as a salary system for the nation's public service personnel. Since then it has spread gradually to the private enterprises (Nippon Steel 1981a, Yawata: 652). This job and capacity rate is determined by ranking employees to a detailed grading classification, which is made by reference to job execution qualification, with the assessment of age and length of service as the basic criteria. This rate was introduced in the early part of the 1960s and by 1975 up to 51.3 per cent of clerical employees and 29.3 per cent of manual employees (in the case of Nippon Steel Corporation) were being paid under this system (Imada 1981: 212). There were two reasons for this development. One was that the existing salary system could not cope with the rise in initial salary levels and they were forced to modify the system based on seniority rate and to restrain the increases in the total salary cost. The other was that the new calculation, which includes personnel assessment, allows greater personnel control.

The efficiency rate is a group efficiency earning applied to manual workers and the percentage allocation of this rate decreased from 29.1 per cent of the wages received in 1967 to 9.7 per cent in 1977, due to plant and equipment modernisation and automation (Tekkō Renmei 1981: 546). In addition to the above salary elements there are various allowances which account for only a few per cent, namely family allowance, shift allowance, special duty allowance, housing allowance, transportation allowance, meal allowance, diligence allowance, education allowance, overtime allowance and night allowance. Managerial employees, such as chiefs at sub-section, section and department level, are given managerial position allowances and the amount of seasonal bonuses (twice a year) becomes larger with increase in rank.

The characteristics of the wage/salary and promotion systems practised in the BSC can be shown in the following three points. First, wages and salaries have not been as high as in other industries. Although in 1963–73 they stood at an overall level of around 102 to 105 per cent of the manufacturing industry average, the iron and steel industry was in third place among manufacturing industries during the 1960s, and dropped to fourth in the 1970s, after automobiles, paper and printing and oil refining (Tekkō Renmei 1974a: 281).

Second, a wage/salary is given as a reward for work in a particular occupation or job, and therefore in every wage/salary the percentage of the basic rate for a specific job in a specific occupation accounts for 90–95 per cent of the total. The wage/salary reaches its highest proportion when the manual employees are in their twenties. For example, the increase in the wage of operatives is achieved when they move with the seniority principle to a higher job on the same promotion line. The level of wages for junior operatives under 21 years old is as low as 34.5 per cent of the average wage of all operatives (1972).[10] Maintenance and craftspeople are divided up into skilled, semi-skilled, other workers and apprentices. In comparison with the wage level of skilled workers, in 1972 semi-skilled workers stood at 88.8 per cent, other workers at 79.2 per cent and apprentices at 38.5 per cent.[11] The skilled workers account for about half of the maintenance and craftspeople group and they are composed of many occupations with different wages. Ancillary workers also include many occupations and the wage differs from occupation to occupation.

Salaries for managerial, technical and clerical employees are graded in each occupation and job. If a worker remains at a certain position and job, salary increments will be maintained for up to six years, depending on length of service, but will thereafter cease to rise. The level of salary for junior staff (16–19 years old) is 50–60 per cent that of employees in the same group at 20 years old.[12] Supervisory employees are also classified into specific occupations and salaries are determined by those occupations. Even if staying in the same supervisory job of the same occupation, the salary increases for up to four years, but will not increase after that. Jobs of middle management and management are all classified and salaries are all determined by them. Within the same job the salary is further graded, thus making salary increase possible throughout the length of service. The salary of senior management is determined by individual negotiation with the director of Personnel Development and Services.

As reviewed above, the wage, salary and promotion system in both industries reflect the difference in the importance of the steel industry in each economy as well as the difference in the basic concept of employment. The existing wage, salary and promotion system which integrates factors of the management system in order to create a system compatibility is thus a reflection of control in business organisation, but at the same time is a condition for managerial strategy.

CORPORATE INDUSTRIAL RELATIONS

The nature of industrial relations is formed in the process of industrialisation and the steel industry is a typical case in which we can identify national diversities. There are two major characteristics of industrial relations in the Japanese large steel companies. First, there exists an enterprise union, or one union which stands for a single whole enterprise. The formation of enterprise unions began when labour unions after the war were organised at each establishment and enterprise. There was a strong post-war impetus towards industrial unions, but this was prevented by the suppression of the General Strike called in 1949, the extinction of the right to strike for government and public workers in 1948, the Dodge Line[13] in 1949 and the Red Purge in 1950 (Hasegawa 1993), all of which paved the way for the successful emergence of enterprise-based unions (Takahashi 1978: 37).

In the case of the Nippon Steel Corporation, a total of twelve unions, each of which exists as a unit at each works, together form an enterprise union called the Nippon Steel Corporation Trade Unions Federation.[14] Both blue- and white-collar workers are organised in the same union and membership is limited to regular employees of the Nippon Steel Corporation, i.e. excluding sub-contract and temporary workers. In this way the equality and homogeneous nature of the union members within the enterprise is assured.

An enterprise union makes it basic policy to improve the working conditions of employees in economic terms, and supports plant and equipment modernisation and managerial rationalisation on the premise of fair distribution of profits (Nippon Steel Kimitsu Labour Union 1980: 887). In such an organisation, as inter-enterprise competition becomes harder, the 'our' or 'us' and the 'collective' consciousness in the unit enterprise are apt to be strengthened at the expense of horizontal class consciousness (Hazama 1974: 22, 247). The 'our' or 'us' consciousness realises the integration of the individual and the whole by restraining individuality. Thus the form of trade union organisation becomes collective and centralised, making it difficult to establish labour–management negotiations at the workshop level.

Second, the foremen who are at the forefront of management are the core of the workshop community. They are not union members and their area of supervision includes fewer than ten workers. Workshop foremen are almost always the oldest and their length of service is the longest. The position is the highest rank that manual workers can generally aspire to and to reach this level they must pass an extremely difficult elimination process. The crucial quality required of a foreman is the ability to lead the workshop community. The foremen are said to be trusted and respected by workers at the workshop and their position resembles that of a father figure. The job breaks down into components of labour management (16.4 per cent), education and training (12.6 per cent), *Jishukanri Katsudō* (self-management activity) (10.4 per cent), work management (30.9 per cent), safety management (13.1 per cent), cost control (10.3 per cent) and others (7.5 per cent) (Tekkō Renmei 1973, *Tekkō no IE*: 38). Labour management, which takes up a relatively high proportion, includes maintaining discipline in the workshop, keeping up morale, fostering positive human relationships, controlling the service of workers, making primary personnel assessment, allocating workers, and paying attention to the health of workers (Nippon Steel 1981a, Yawata: 649).

Under the foremen, there are group leaders, who are likely to become foremen in the future and they are in general branch officers of the enterprise union. Thus labour management at the workshop level creates a firm foundation of labour–management co-operation with the foreman as the core of management control, acting with support from the branch of the trade union.

One of the characteristics of industrial relations in British steel is a large number of trade unions in the one enterprise. In the British Steel Corporation there were nineteen trades unions: two industrial unions (ISTC being the largest of all the trades unions in the BSC), nine craft unions, five general unions, two staff and management unions and one management union. Although the existence of such a large number of unions reflects the history of unionisation of British workers during the process of development of British capitalism, one common feature of all of them is that they are organised horizontally. These horizontal unions, possessing a strong class orientation fostered throughout their long industrial history, organise as many as 95 per cent of all employees. Under such a situation, although industrial relations in the iron and steel industry are said to be much better than in other industries, the workers' 'them and us' consciousness is extremely strong. This hinders any realisation of industrial relations with a sense of community (Hazama 1974: 61–71). The workshop is, rather than a place of community, a place of conflict between management and labour, and in addition it is the place to protect workers' own interests by identifying the demarcation between unions. As the TUC has a rather weak influence over its member unions (although it is a joint association of trades unions) and each union's central control power is also rather weak, conditions have been created for a large number of unofficial strikes.[15]

A second characteristic is that foremen are mainly occupied in organising and controlling the flow of work and oversee a comparatively large number of people, so their personnel management function over the workers is very weak or non-existent.[16] Foremen are not always the oldest and they are not chosen because of their long length of service. Vacancies for foremen positions are usually announced in the Corporation newspaper and then applicants will have an interview with a group of managers, including a representative from the personnel function of the works.[17] At the Scunthorpe works the selection of foremen is done by interviewing those who have finished a course of supervisory education and training. University graduates who are in the period of training are encouraged to become foremen as well.[18] The salary of foremen is not that high in comparison with other workers and the position is not as authoritative and attractive as it is in Japan.

Foremen are in general members of a union in order to protect their own interests, but they place themselves on the side of management, drawing a line between them and general workers. Therefore they are certainly supervisors over the workers at the workshop level, but are not their leaders in terms of human relationships. Generally workers establish and maintain their workshop group and human relationships through the membership of trades unions (Dore 1973: 250–1).

Thus industrial relations, once established in each country, have formed an important part of the corporate environment and react to managerial strategy, in particular in the use of workforce, and to technological innovation. It would appear to be more advantageous for management to have in-enterprise unions, as these can be accommodated easily to their business objectives.

CONCLUSION

The institutional circumstances reviewed here have been treated as internal corporate environments for the study of technological innovation at enterprise level. The author, however, believes that the institutional arrangement should be regarded both as a cause and effect in relation to the growth of the enterprise, interacted by managerial strategy and social relationships at work within the business organisation. The differences identified in the broad range of practices established in the respective industries may be ascribed, therefore, to the differing stages and processes of development found in each.

In sum, it can be concluded that the institutional practices that have arisen in Japan have been highly appropriate, becoming both cause and effect of the long-term, large-scale technological progress which took place up to 1974. On the other hand, the practices that developed in Britain and have been maintained with some modifications appear not only to have been unsuitable but at times a negative factor for any Japanese-style input of new technology.

Part II

The Japanese steel industry

5 The Japanese steel industry

In the previous chapter we looked at the differences in corporate environment of the two steel industries and assumed that such differences might affect the mode and result of their modernisation.

The Japanese steel industry has established a new production system, characterised by a combination of advanced technology and a unique approach to the division of labour. The dual structure of the workforce and the optimum automation of production systems is an outcome of managerial strategy as well as a preconditioning factor in Japan's 'co-operative' industrial relations. This chapter will look at further aspects of the modernisation programme.

CAPITAL INVESTMENT

Relative importance of steel

A high level of investment was made by the steel industry in the years following the war, indicating its relative importance in the revival of Japanese industrialism. A total of ¥849.6 billion was spent on plant and equipment between 1945 and 1975, when the industry stepped down its quantitative expansion. Investment during the third rationalisation plan, which began in 1960 and extended over the next fifteen years, greatly surpassed that of any other industry and on average it was 32.6 per cent of the total investment of all manufacturing industries (see Figure 5.1). The subsequent scaling-down is shown, however, in that by 1990 this relative figure had fallen to 13 per cent; the steel industry's paramount position had been supplanted by the electronic-electrical and automobile industries, with 32.4 and 30.9 per cent of total investment respectively (Tekkō Renmei, *Tōkei Yōran* 1992).

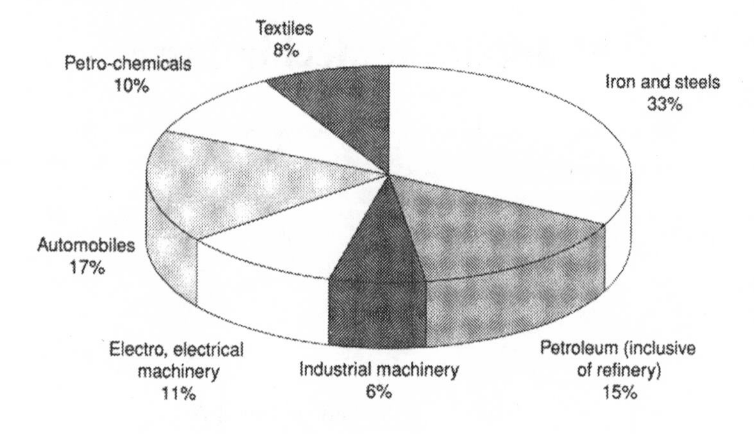

Figure 5.1 Capital investment by major manufacturing industries (1960–75)
Source: Tsūsanshō 1970, 1978

Four macro-economic factors led to the huge capital investment of the steel industry.[1]

1 the rapid increase in domestic demand created by the growth of the construction, shipbuilding, automobile, machinery and petro-chemical industries;
2 the rapid increase of steel exports, mainly to the US;
3 the strong competitive pressures among big steel companies;
4 the government's consistent industrial commitment to steel.

Table 5.1 outlines the post-war development of the Japanese steel industry manifested in the increase in capital investment and its relationship to plant and equipment modernisation. Capital investment increased markedly after 1956, until investment in the latter half of the 1960s was as much as 2.2 times that of the first half of the decade. The growth rate during this period indicates that the primary objective of capital investment was the expansion of steel production capacity, although many hot and cold strip mills were also installed. The number of continuous casters installed increased rapidly from the late 1960s throughout the 1970s.

Modernisation of plant and equipment was not just a matter of more equipment, but has always been accompanied by a high level of automation, involving the comprehensive use of computers throughout the main processes of production. In the late 1960s, a number of new

Table 5.1 Modernisation of the Japanese steel industry

Modernisation Period	*Investment expenditure (100 million yen)*	*Crude steel production (million MT)*	*Crude steel capacity (million MT)*	*Units of equipment*					*Number of employees (1,000)*
				Blast furnace	*LD converter*	*Hot strip mill*	*Cold strip mill*	*Continuous casters*	
Revival period 1946–50	137	4.83	16.17	37	6	1	2	—	164
First modernisation period 1951–5	1,282	9.40	11.27	33	7	3	7	—	212
Second modernisation period 1956–60	6,255	22.13	27.33	34	13	7	26	—	305
Third modernisation period (1) 1961–5	10,138	41.16	53.25	49	45	13	48	5	346
Third modernisation period (2) 1966–70	22,426	93.32	114.63	64	83	19	64	40	382
1971–5	40,258	102.31	152.01	69	98	23	71	122	475
1976–80	37,902	101.61	158.72	65	94	N/A	N/A	141	383
1981–5	40,317	103.75	152.36	54	85	N/A	N/A	154	353
1986–90	32,115	111.00	115.06	46	72	N/A	N/A	148	306

Sources: Tekkō Renmei 1959: 698–9; Tsūsanshō 1973: 78–9; Keizai Kikakuchō 1977: 156; Tekkō Renmei, Tōkei Yōran 1961, 1965, 1970, 1971, 1977, 1986, 1992; Tekkō Kyōkai 1972: 698; Tsūsanshō 1978: 78–9; Tekkō Renmei, *Nihon no Tekkogyō* 1979: 20–3; Tekkō Rōren, *Rōdō Handbook* 1989

Notes: (a) Investment expenditure indicates the total amount of investment used in each modernisation period based on the statistics made public by the Ministry of International Trade and Industry

(b) Crude steel production and the number of employees indicate the figure in the last year of each period

(c) Crude steel capacity and units of equipment indicate the capacity and the number of units as of December of the last year of each period, except for the revival period, which is as of December 1949

integrated steel works were established, raising steel production capacity to 2.2 times that of the first half of the decade. Actual output also rose considerably in this period, but not to the same extent as capacity. The gap between capacity and production widened after 1965, and as explained in Chapter 2, the post-1973 slowdown further emphasised the need to develop exports to counteract the shortfall. Another important point to note is the clear termination of quantitative expansion in the whole industry after 1975, with the emphasis of capital investment shifting to down-stream processes relevant to improving the quality of steel products.

Sources of capital investment

Investigation of the sources of the invested capital will identify the agencies which contributed to steel's modernisation. During the revival period (1945–50), these sources were the Fukko Kinyu Kōkō (Reconstruction Finance Corporation), 14.7 per cent; the Mikaeri Shikin (US Aid Counterpart Fund), 16.1 per cent; shares, 15.0 per cent; debenture stock, 17.8 per cent; and bank loans, 36.3 per cent. It is assumed that the Reconstruction Finance Corporation and the US Aid Counterpart Fund promoted the collection of funds from other sources (Tekkō Renmei 1969: 74).

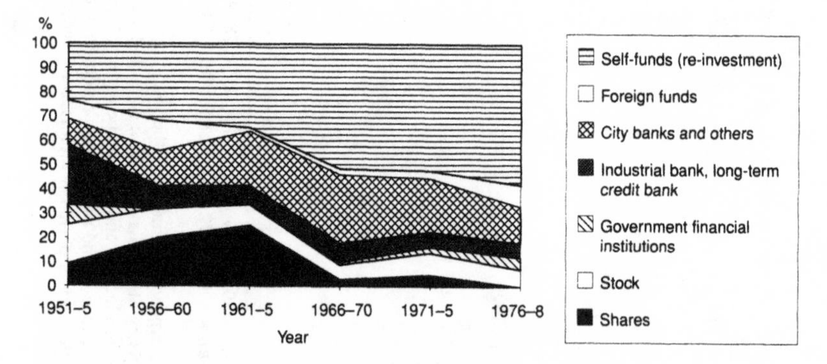

Figure 5.2 Changes in source of investment funds
Sources: Tekkō Renmei 1959: 128–9; Tekkō Renmei 1969: 480–1; Tsūsanshō 1966–78

Figure 5.2 shows some characteristics of the fundraising for plant and equipment investment following the First Rationalisation Period (1951–5), *viz* (Masaki 1968):

First, the percentage of funds raised by ordinary share issue was highest during 1956–65 and since then has decreased dramatically.

Second, the percentage of funds raised by the issue of debenture stock has remained at less than 10 per cent since the 1960s, partly due to the stagnation of debenture markets.

Third, government financial institutions played an important role during the revival and first rationalisation periods, but since 1970, due to increased loans from the Japan Development Bank and the Japan Import and Export Bank, the involvement of government institutions has increased only modestly.

Fourth, the percentage of loans from the Industrial Bank of Japan Ltd and the Long-Term Credit Bank of Japan was quite high during the first rationalisation period, but fell thereafter to below 10 per cent and is still decreasing. Nonetheless, financial relations between these two institutions and the large steel corporations are very strong. For example, in 1975 the Industrial Bank of Japan ranked highest in terms of loans to and share ownership in Nippon Steel, third in loans and eleventh in share ownership in Nippon Kōkan, third in loans and fifth in share ownership in Sumitomo Metal Industries, and highest in loans and seventh in share ownership in Kobe Steel. Meanwhile, the Long-Term Credit Bank of Japan ranked highest in loans and second in share ownership in Kawasaki Steel (Tōyō Keizai 1975: 139–40).

Fifth, the percentage of loans by city banks and others was over 20 per cent during the 1960s, but during the three years 1976–8 generally decreased. Nippon Steel has kept about the same degree of financial relationship with the various main city banks; Kawasaki Steel has much stronger financial ties with the Industrial Bank of Japan and the Long-Term Credit Bank of Japan than with the main city banks; Nippon Kōkan has stronger ties with the Fuji financial group than with other city banks, while the Fuji financial group ranks highest in both loans to and share ownership in Nippon Kōkan; Sumitomo Metal Industries has closest ties with the Sumitomo financial group, while the Sumitomo financial group ranks highest in loans to and share ownership in Sumitomo Metal Industries; Kobe Steel has its strongest ties with the Sanwa financial group, which ranks highest in loans and share ownership in Kobe Steel (Tōyō Keizai 1975).

Sixth, the percentage of foreign funds, supplied mostly by the World Bank and the Import and Export Bank, was highest during the second rationalisation plan. During the first half of the third rationalisation period it declined considerably to 1.3 per cent, but since then has gradually increased. It is important to note that the percentage of foreign funds in the total long-term debts of major steel corporations is quite high. For example, Nippon Steel has debts amounting to 12.7 per cent of the total long-term debt with 45 foreign banks; Kawasaki Steel has 12.9 per cent, with 52 foreign banks; Nippon Kōkan 11.5 per cent, with 52 foreign banks; Sumitomo Metal Industries 14.5 per cent, with 53 foreign banks; and Kobe Steel takes up 15.3 per cent, with 57 foreign banks (Major Steel Companies, Yukashōken Hōkokusho 1977). Of these overseas institutions, those belonging to US financial groups are the most important and influential.

Seventh, the percentage of self-funding, of which the greatest proportion is from depreciation funds, is increasing steadily; since the late 1960s it has run at over 50 per cent. It is evident that various measures to increase depreciation funds made them available for investment (Tekkō Renmei 1969: 489), but more fundamentally the increase in self-funding means that capital formation has been achieved quickly and more easily. This is shown by the changes in the amount of various value-added components at Yawata Steel between the years 1955 and 1969. There, the percentage of personnel expenditure declined, while the net profit and depreciation fund increased steadily (Nippon Steel 1969).

Thus, up to 1965 more than half the total of investment funds was secured from outside sources, in particular from shares and city banks; but the situation has since shifted to self-funding and greater independence in securing capital investment as the corporations have moved to a higher stage of development. The insignificant role of shares in capital investment is also noteworthy.

DEVELOPMENT OF MODERNISATION

Modernisation in the reconstruction period (1945–50)

The war caused relatively little damage to steel production equipment, but only three blast furnaces, of a total capacity of 418,000 tonnes, were in operation in the year after hostilities ceased, and only twenty-two open furnaces, totalling 677,000 tonnes capacity. However, soon after the end of the war the US occupation policy began to change; economic aid from the USA was taken advantage of and US economic advisors helped develop a plan for the rationalisation of the steel industry. Initially, the rationalisation of plant and equipment during the reconstruction period (1946–50) was mainly concerned with repair of already existing blast furnaces and installation of new rolling equipment. Blast furnaces were repaired and improved in twenty-one cases, and newly established in two cases; similar figures for open hearth furnaces were ten and five, and for rolling mills two and 126. In addition, fifty-seven electric furnaces were repaired and improved (Tekkō Renmei 1969: 689).

Equipment rationalisation during this period was affected by US policy during its post-war occupation of Japan, but companies took advantage of the change of attitude when the strict Pauley interim report of December 1945 ('steelmaking capacity which will exceed 2.5 million tonnes shall all be designated as reparation') (Tekkō Renmei 1959: 6), was superseded by the Strike Report of March 1948, which looked forward to the reconstruction and development of the Japanese steel industry.

The first modernisation period (1951–5)

The first modernisation plan was prepared by the Industrial Rationalisation Council and approved by the Japanese cabinet. The capital expenditure for this rationalisation plan amounted to ¥128.2 billion, 50 per cent of which went towards the rolling processes, 12.6 per cent towards ironmaking, 10.7 per cent towards steelmaking and 26.7 per cent towards other processes (Tekkō Renmei 1959: 697). The emphasis was on repair, replacing obsolete equipment and bringing rolling equipment up to current standards. Of this total, 85.1 per cent was spent by the six major corporations of the time. Steel output and capacity in the final year of this plan did not differ much from 1950. While the number of blast furnaces decreased, hot and cold strip mills increased, and other rolling and surface treatment equipment was modernised.

The second modernisation period (1956–60)

The total capital expenditure for the second modernisation plan amounted to ¥625.5 billion, which was about five times that of its predecessor. The ambitiousness of this plan derives from a sudden rise in the export of steel products from the autumn of 1954, restrictions on the import of scrap iron from the United States, and increased reserves of funds in the major corporations. This rationalisation plan effected the modernisation of existing works and expansion into adjacent areas to establish new integrated works. The specific programmes carried out by the major corporations at this time illustrate the mood of this period.[2]

1 Yawata Steel Corporation, Tobata District (Yawata Works): The expansion of this district began in 1955 as the world's first steelworks on land reclaimed from the sea. The No. 1 blast furnace (1,603 m^3)

started operation in September 1959. This works, in expectation of an increase in demand for sheet steel, put its emphasis on the production of hot rolled sheet, cold rolled sheet, tin plate, galvanised sheets and spiral tubes.

2 Nippon Kōkan KK, Mizue District (Keihin Works): This company installed new blooming and slabbing mills, hot strip and cold strip mills in 1959. The No. 1 blast furnace began operation in 1962. This works specialises in the production of sheet, jobbing plate, tin plate, galvanised sheets, colour-coated sheets and pre-painted galvanised sheets.
3 Kobe Steel Ltd, Nadahama District (Kobe Works): Reclamation began in May 1957 and the No. 1 blast furnace started operation in 1959. By June 1962, the No. 2 blast furnace, LD plant, No. 3 blooming plant, No. 4 and No. 5 wire rod plants and small sections plants had come on line. This is the largest works in the world specialising in the production of wire rods and steel bars.
4 Sumitomo Metal Industries Ltd, Wakayama Works: 1957 saw the commencement of work on this site as an integrated iron and steel works. The No. 1 blast furnace started operation in 1961, No. 2 in 1963 and No. 3 in 1965. This location specialises in the production of various tubes and surface treated sheets.

During this modernisation period, advances in rolling production were achieved, as seen in the growing number of strip mills, which reflected an increase in sheet steel demand caused by the expansion of the consumer durables market. A greater need was seen, however, for the modernisation of iron and steelmaking processes and this led to eleven blast furnaces and six LD converters being established. The last year of this period celebrated a steel output of 23 million tonnes, about 2.3 times more than that of 1955, the final year of the first modernisation period (1955). Steelmaking capacity also increased to 28 million tonnes, about 2.5 times more than that of 1955.

The third modernisation plan (1961–5/1966–70)

During the first half of the third modernisation period, annual economic growth hit an average of over 10 per cent, and the domestic demand for steel products expanded together with further increases in export demand. The total capital expenditure during this period amounted to ¥1,013.8 billion, though the rate of investment in iron, steel and rolling departments as a proportion of the total declined slightly in comparison with the second rationalisation plan.

A remarkable increase in investment was apparent in the industry-related facilities necessary for building large-scale and up-to-date integrated works. Among various modernisation projects in this period, the most significant was the construction of these new integrated steelworks as centrepieces for industrial complexes. The idea of the complex was to enable the government's Income Doubling Programme, which aimed at speedy economic growth. Nippon Steel Corporation's works at Nagoya (run before amalgamation by the Fuji Steel Corporation), Sakai and Kimitsu (formerly of the Yawata Steel Corporation), Nippon Kōkan KK's Fukuyama works, and Kawasaki Steel's Mizushima works were such ambitious new industrial plants. LD converters and many large high-speed strip mills were installed at these sites and in 1965 the overall number of blast furnaces stood at forty-nine, LD converters at forty-five, hot strip mills at forty-eight and continuous casters at five. The modernisation and rationalisation programmes led to steel output in this year of around 41 million tonnes amid a steelmaking capacity of about 53 million tonnes. The second half of this modernisation drive, from 1966–70, involved approximately twice the expenditure on plant and equipment as the previous five years, amounting to ¥2,341.1 billion. Of this, 13.8 per cent was spent on new works construction, 72.5 per cent on the new works initiated in the previous years and 13.7 per cent on general maintenance and repair; thus, the bulk of investment was in ongoing projects. New construction programmes included three integrated works: Kakogawa, for Kobe Steel Ltd, Kashima for Sumitomo Metal Industries Ltd, and Oita for Nippon Steel Corporation.

The features of modernisation in this period were the introduction of comprehensive on-line computer systems at the Fukuyama works of Nippon Kōkan KK, the Mizushima works of Kawasaki Steel, the Kashima works of Sumitomo Metal Industries and the Kakogawa works of Kobe Steel, and the introduction of continuous casters in large numbers by the main steel corporations. At the same time the number of LD converters and cold strip mills continued to increase. As a result of such modernisation, in 1970, the last year of this rationalisation period, there were sixty-four blast furnaces, eighty-three LD converters, sixty-four hot strip mills and over forty continuous casters. Steel output in 1970 was about 93 million tonnes and steelmaking capacity was increased to 114 million tonnes, about 2.15 times higher than in 1965.

Modernisation in the 1970s (1971–5)

The next phase of modernisation was carried out in a different economic atmosphere from previous periods, because of the termination of high economic growth which followed the October 1973 oil crisis. The overcapacity of the steel industry became apparent in these difficult economic conditions and economic policies to promote the export of goods and capital were adopted, along with the issue of more public bonds. Workforce reduction, mergers and amalgamations of small- and medium-sized firms became familiar events. Yet despite the recession, the steel industry in this period again doubled its expenditure on plant and equipment, spending ¥4,216.3 billion. This broke down into 9.6 per cent on new works construction, 71.6 per cent on ongoing projects, and 18.8 per cent on maintenance and repair.

Thus, investment was concentrated on improvements, maintenance, and repairs of existing equipment. Various features of this period can be identified, *viz*: first, many continuous casters were installed in medium-sized enterprises with electric and/or open hearth furnaces; second, an increase in volume capacity was achieved despite the number of blast furnaces and LD converters remaining static; third, the degree of automation and areas of its application expanded further; fourth, the rate of investment spent on safety measures increased, accounting for 18.6 per cent in 1975.

Major expansion and modernisation continued in the Japanese steel industry until 1973, shifting over time to a consolidation of its situation, represented by maintenance and partial improvement of what had already been achieved. Overall, the establishment of large-scale, high-speed, continuous and automated production systems had become a general characteristic of the modernisation of the Japanese steel industry.

TECHNOLOGY IMPORT AND R & D

Development of technology import

During, and for a while after, the war, no technology was imported from other countries. However, the Foreign Exchange and Trade Control Act of 1949 and the Foreign Investment Law of 1950 reopened the way for the importation of technology. In 1968 there was substantial liberalisation of the technology import regulations and they were thoroughly liberalised by 1974. The steel industry was not neglected in the promotion of such imports and by 1975 508 Class A and 406 Class B technology items had been brought into the country. Figure 5.3 shows the trend of technology import during 1950–73. The largest importation occurred between 1965 and 1970, amounting to 218 imports of both A and B class items.

Breakdown into country of origin reveals a strong dependence on American technology, accounting for 52.9 per cent of Class A (*Kō-shu*) technology and 50.3 per cent of Class B (*Otsu-shu*).[3] By comparison, West Germany supplied 18.6 per cent of Class A imports, Switzerland 6.1 per cent, the USSR 4.3 per cent and France 3.5 per cent. In Class B, West Germany supplied 37.2 per cent and France 8.4 per cent.

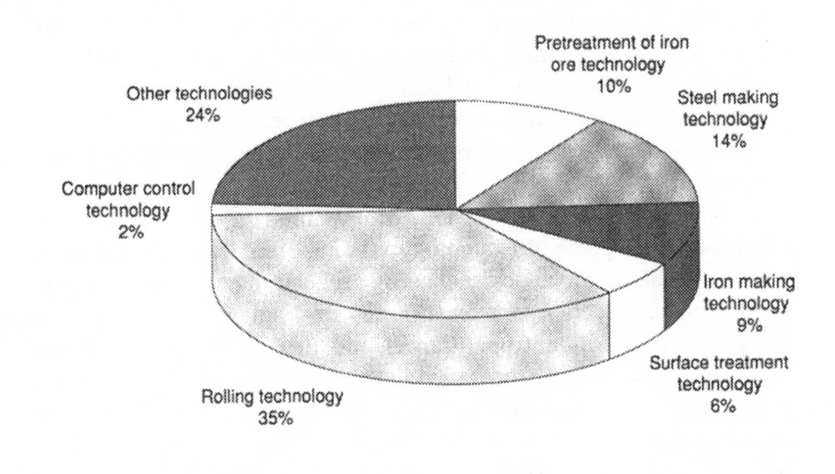

Figure 5.3 Number and types of imported technologies (1950–75)
Sources: Jyukagakukogyo Tsūshinsha 1974; Tekkō Shimbun-sha 1975, 1976

Figure 5.3 shows the number of important technology imports of the steel industry classified according to technological application. Rolling technology imports are the highest (188), with steelmaking ranking second at seventy-six items. The number of imports grouped in 'other technologies' is also high, because it includes technologies which are not immediately related to iron and steel production, i.e. shipbuilding and heavy machine production technologies used by Nippon Kōkan KK and Kobe Steel Ltd, two of the five major corporations surveyed.

The most active import programme by the five major corporations was in the ten years from 1955 to 1965; 1960–5 were the peak years. Over the whole period from 1950–75, the five major corporations accounted for 58.4 per cent of the total important technology imports, but the intensity of import in the first fifteen years of this period was much higher than thereafter. The decline was such that in 1970–5 the share of imports was only 24 per cent.

This trend indicates that by 1965 the 'big five' corporations had completed their import of the best available technology required to establish internationally competitive steel works; this then freed them to build up their own research and development activities, based upon the imported technology. The decline of technology import and the beginning and increase of technology export by the five major corporations all occurred around the same time.[4]

The lower level of technology import by the major corporations can be explained also by the post-1965 increase of technology import by small- and medium-sized producers. These concerns are related in varying degrees to the major corporations by subsidiary contracts and were under pressure to catch up with the technology used in their parent companies.

It is also noteworthy that the type of technology required determined which country a piece of technology was imported from. For example, in the ironmaking processes, technology related to coke and the repair, design and construction of blast furnaces are from West Germany; that of heavy oil injection from France; and that of high top pressure operation from the United States and the USSR. In the steelmaking process, the soaking pit technology came from the United States, the oxygen top-blown converter and vacuum refining process from West Germany and Austria, continuous casting technology from the United States, Switzerland, West Germany, the USSR and France, and new steel product manufacturing know-how and alloy production technologies from the United States. Most of the technology relative to rolling and surface treatment as well as computer control was from the United States.

Factors for technology import

Several factors promoted such substantial technology import. The increasing competition among firms during the high economic growth period was one of these. One of the management goals common to all the major Japanese corporations was to achieve plant and equipment modernisation and increase market shares. The growing demand for steel in the burgeoning manufacturing industries meant that if plant and equipment were not updated, demand could not be met either in the required quality or quantity. Expectation of the elimination of protective measures, which in fact occurred in 1967 at the Kennedy Round, also obliged them to adopt the most modern equipment in order to become competitive in product quality, because the West was still superior in some product areas, even though the Japanese companies were competitive in price.

The main reason behind investment in new facilities and equipment was a significant expansion of demand for steel products, especially in the construction, home and manufacturing, shipbuilding and automobile industries during 1960–75. The demand grew as much as three- to five-fold, depending on the industry: the total amount of orders for ordinary steel products in 1960 was 11.28 million tonnes, increasing to 46.15 million tonnes in 1975 (Tekkō Renmei, Tōkei Yōran 1970, 1976). Competition was intensified by both the rapid growth in demand and the liberalisation of trade and foreign exchange which began in 1960. The steel industry advocated the necessity of plant and equipment modernisation to establish mass production systems by saying,

> Increase in competitive power is absolutely necessary not only to compete with the foreign steel products imported with the liberalisation of foreign trade, but also to increase the volume of steel export in addition to supplying the domestic market with steel products at lower prices than the international market.
>
> (Tekkō Renmei 1969: 785)

Second, pressures to raise wages and improve labour conditions caused the promotion of technology import. Any increase in wages and salaries or improvements in labour conditions add to costs in both relative and absolute terms, which then puts pressure on companies to modernise plant and equipment and promote technical improvements, expanding investment in plant and equipment. Modernisation of plant and equipment and technical improvement were particularly hastened both by the shortage of young workers which became apparent in 1960, and the increase in labour expenses caused by the rise of wages and improvement of working conditions, themselves mainly created by the labour shortage and the efforts of the labour movement. If we set labour expenses incurred for an average worker in one of the five major corporations in 1960 at 100, it rose to 242 in 1970. For those Japanese steel firms which had to depend on imported raw materials to a large extent and consequently had to pay higher interest rates, such a rise in the index for labour expenses surely necessitated plant and equipment modernisation and technical improvement.

R & D investment

The basic technology of iron and steelmaking has not changed in principle during the post-war period, but the improvement and development of systems of combined technology and the comprehensive use of computers have established a large, continuous, high-speed and automated production system. In Japan the R & D expenditure rate of this industry accounts for only 4.6 per cent of the total R & D expenditures of all industries, and ranks fifth, behind the electric machinery industry (26.2 per cent), the chemical industry (18.3 per cent), the transport machinery industry (16.5 per cent) and the machinery industry (6.7 per cent) (Kagaku Gijyutsuchō 1975: 226).

Investment on R & D began to increase remarkably from 1960 when the third rationalisation plan started. As shown in Figure 5.4, the total R & D expenditure of the major six corporations was ¥3.8 billion in 1959, while in the following year it increased to ¥6.1 billion, 1.6 times more than the previous year. The amount of R & D expenditure of the steel industry as a whole increased 19.2 times, and that of the major six corporations 11.3 times, from 1960–74. Periodic observation of the rate of increase of R & D expenditure shows that from the latter half of 1960 it increased at a rate of more than 20 per cent and peaked at 45 and 51 per cent in 1973 and 1974. The ratio of R & D expenditure to sales has also increased. The number of personnel engaged in R & D activities increased 2.8 times from 1960 to 1974.[5]

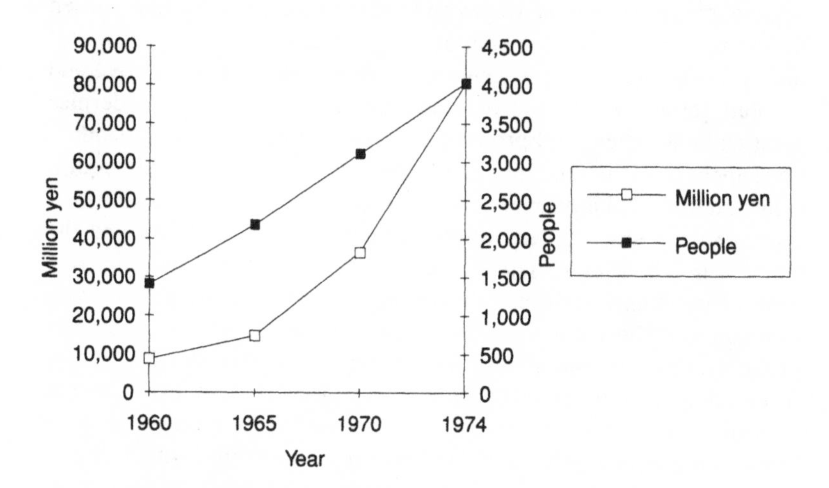

Figure 5.4 R & D activities of steel industry
Sources: Sorifu Tōkeikyoku 1965, 1966, 1969, 1975, 1976; Tekkō Shimbun-sha 1969, 1970, 1976

Development of R & D organisation

Before 1958, when R & D expenditure was small, there existed no formal R & D organisation in any corporation. Emphasis was placed instead on the improvement and modernisation of existing equipment at each works. But from 1960, when expenditure on R & D began to increase rapidly, we begin to see the establishment of R & D departments. It is possible to divide this trend into two periods.

The first started around 1960, when each major corporation established a central R & D institute at head office. The first to be established was the Technical Research Institute at Kawasaki Steel Corporation in August 1957. In April 1958, Yawata Steel Corporation followed with the Tokyo Research Institute, while in August that year Sumitomo Metal Industries set up their Central Technical Research Institute. In 1960, Nippōn Kokan KK established their Research Institute, which is directly under the control of the head office, and Kobe Steel also set up its Central Research Institute that year. These institutes were set up at the start of the third rationalisation programme to prepare for the liberalisation of foreign trade as well as to cope with increasing domestic competitive pressure. In order to cope with these two situations, corporations were forced to introduce numerous new technologies from abroad as fast as possible. To optimise the value of these, it was necessary to establish R & D institutes to examine the imported technology with a view to adapting it to the existing production methods and equipment. Sometimes the introduction of new technology also required change and improvement in the home equipment. Furthermore, the strong demand for new types of steel products promoted research activities in the major corporations, thus encouraging the establishment of R & D institutes.

The second period, from the latter half of the 1960s to 1970, witnessed substantial development in these R & D organisations. Sub-organisations, originally set up for specific research projects, were reorganised and the whole system incorporated into head R & D departments, and this is the situation we see today.

There are two types of R & D organisations in the major corporations. The first is found in corporations which have one (Nippon Steel Corporation and Kobe Steel Ltd) or more (Kawasaki Steel Corporation, Sumitomo Metal Industries Ltd and Nippon Kōkan KK) research institutes. The second type is research and development in individual steelworks, incorporated into a central R & D department (Nippon Kōkan KK and Kawasaki Steel Corporation) or independent (Kobe Steel Ltd, Sumitomo Metal Industries Ltd and Nippon Steel Corporation).

MODERNISATION IN THE MAIN PRODUCTION PROCESSES

Japanese companies have established several extensive steelworks on coastal sites, taking full advantage of official economic policies for rapid economic growth: among these measures are the use of public finance funds, special tax reductions, low rate finance, and regional development policies inviting steel enterprises to establish works. Including extended existing sites, there were fourteen of these new steelworks (see Table 5.2). These modern integrated works are the key production units of their companies, and naturally the more recently established ones are even more modern in their facilities and equipment. The main equipment in the ironmaking, steelmaking and rolling processes is large in scale, continuous, high-speed and automated; and the bulk production of iron and steel is well combined with mass production rolling systems.

Modernisation of plant and equipment, when considered according to the flow of materials, is logically realised with ironmaking first, followed by steelmaking and then rolling. In reality, however, it did not progress in that order, but rather developed in a kind of reciprocal activity between processes. A principal factor encouraging the enlargement of blast furnaces was the abolition of open hearth furnaces due to the full-scale adoption of LD converters. The use of continuous casters has in turn been promoted by the adoption of LD converters, their rapid enlargement and operational efficiency, and also by the full adoption of wide-range hot strip mills (Tekkō Kyōkai 1974; 1975; 1976; 1977).

Ironmaking process

The main equipment in this process is the blast furnace. The biggest blast furnace that existed pre-war was 1, 4003 m^3 in its internal volume. The post-war enlargement of blast furnaces began in 1959 with the construction of the No. 1 blast furnace (1,603 m^3) at the Tobata works of the Yawata Steel Corporation. From then on, large-scale blast furnaces continued to be constructed with capacity increasing greatly after 1965. In 1976, the No. 3 blast furnace (5,050 m^3) at the Kashima works of Sumitomo Metal Industries Ltd and the No. 2 blast furnace (5,070 m^3, the largest in the world) at the Oita works of Nippon Steel Corporation began operation. In comparison with the number of blast furnaces over 2,000 m^3 total volume in other countries, in 1977 Japan had thirty-eight furnaces, USSR twenty-nine, West Germany seven, the USA five, Italy five, France four, and the UK one (Tekkō Renmei 1978: 37).

Developments in ironmaking techniques were also achieved along with the enlargement of blast furnaces. They were mainly developments in pre-treatment techniques of iron ore (ore sizing, blending, self-fluxing sinter, etc.); developments in blast furnace operation techniques (oxygenation, high temperature blast, composite blast including heavy

Table 5.2 Outline of newly established steel works and enlarged old steel works

Name of steel works	*Time of the first BF operation*	*Number of BF (max. capacity)*	*Number of LD converters*	*Crude steel production capacity (10,000 MT)*	*Number of employees*	*% of sub-contracting*
Kawasaki Steel: Chiba	1953	5 (2,584)	5	650	11,525	47.2
Kobe Steel:						
Kobe (Nadahama District)	1959	3 (1,845)	3	236	5,822	36.4
Nippon Steel:						
Yawata (Tobata District)	1959	4 (4,140)	8	856	20,142	45.7
Sumitomo Metal: Wakayama	1961	5 (2,700)	7	922	10,154	54.5
Nippon Kokan:						
Keihin (Mizue District)	1962	4 (3,900)	7	664	11,661	33.9
Nippon Steel: Nagoya	1964	3 (3,240)	5	720	8,630	50.1
Nippon Steel: Sakai	1965	2 (2,800)	3	450	3,512	53.8
Nippon Kōkan: Fukuyama	1966	5 (4,617)	8	1,600	11,791	61.1
Kawasaki Steel: Mizushima	1967	4 (4,323)	6	1,200	12,546	50.6
Nippon Steel: Kimitsu	1967	4 (4,930)	5	1,000	7,669	65.7
Kobe Steel: Kakogawa	1970	2 (3,850)	3	630	6,495	45.5
Sumitomo Metal: Kashima	1971	3 (5,050)	5	1,150	6,624	68.5
Nippon Steel: Oita	1972	2 (5,070)	3	800	3,699	70.4
Nippon Kōkan: Ogishima	1978	2 (4,000)	3	600	11,725	37.5

Sources: Major companies, *Yukashōken Hōkokusha* 1977; Tekkō Shimbun-sha 1976; Tekkō Rōren, *Rōdō Handbook* 1976

Note: As Ogishima is part of Keihin Works of Nippon Kōkan the number of employees and the percentage of sub-contracting are those of Keihin Works

oil injection, and high and super high top pressure operation); and developments in computer control.

Steelmaking process

The main modernisation of equipment in this process was the introduction and enlargement of LD converters which replaced open hearth furnaces, and of continuous casting equipment. The LD converter was developed in 1952 in Austria. In Japan, Yawata Steel Corporation introduced the LD converter in 1957 and Nippon Kōkan KK in 1958. After that, steelmaking by LD converters spread quickly as it has a superior performance in many respects over other methods (Nippon Steel 1976: 31). In particular, the shortage of scrap iron and the rapid change of energy source from coal to oil which occurred in the 1950s and the ability to supply industrial oxygen in large amounts have made the introduction of LD converters all the more advantageous. Steelmaking by LD converter came to account for 82.5 per cent of the total steel output, surpassing that of the United States and West Germany by 1975 (Tekkō Renmei, *Tōkei Yōran* 1978).

Enlargement of blast furnaces caused a demand for larger LD converters. While the first LD converter had a capacity of fifty tonnes and the average size until 1960 remained less than 100 tonnes, it grew to 110–80 during 1961–7, then to 220–340 tonnes during 1968–75. The LD converter at Oita, which as we have seen also has the largest blast furnace in the world, has a capacity of 340 tonnes. In a comparison of ownership of large LD converters (more than 201 tonnes/charge) by country, Japan had thirty-two, the USA thirty-one, the USSR eighteen, West Germany fifteen, the UK eight, France seven and Italy six as of 1977 (Tekkō Renmei, *Tōkei Yōran* 1978).

The increased use of LD converters and their enlargement caused the development of various operational technologies:

- automatic measuring
- automatic control of auxiliary material supply equipment
- automatic crane operation
- automation of information systems
- development of environment management (gas treatment technology for basic oxygen converter, auxiliary dust collector, house dust collector)
- development of computer control (control with dynamic model, full automatic control of lance oxygen in blowing)
- extension of the life of the converter

- development of technologies related to the LD converter (desulphurising technology of hot-metal, vacuum refining technology, deoxidisation and alloying technology).

Continuous casting equipment takes the place of ingot-making, soaking, and the blooming and slabbing process. This was developed in West Germany (Mannesmann process), the UK (BISRA process) and Switzerland (Concast process). Sumitomo Metal Industries introduced the Concast equipment in 1955 for the first time in Japan. But it was only after 1970 that this equipment came to be introduced full-scale. The total number of Concast for slabs was seven in 1967 but it increased to thirty-four in 1977 and that of bloom and billet also increased from ten in 1967 to ninety-six in 1977. Today, major corporations have large-scale continuous casting equipment for slabbing and blooming in almost all new integrated works, and many smaller producers with open hearth furnaces and electric furnaces have continuous casting equipment for billet making. Production of semi-finished products by the continuous casting process amounted to 41.81 million tonnes at the end of December 1977 and accounted for 41 per cent of the total semi-finished products. The first continuous caster had a capacity of 10 tonnes/hour but now the largest capacity is 370 tonnes/hour (Keihin Works, Nippon Kōkan KK). In 1972, the Oita works of Nippon Steel Corporation installed all continuous casting equipment (200 tonnes/hour).

Comparing the percentage of steel production by continuous casters by country in 1975, Japan was 31.1 per cent, the USA 9.1 per cent, the USSR 6.9 per cent, West Germany 24.3 per cent, Italy 27.0 per cent, and the UK 8.5 per cent (Toda 1984: 14).

Rolling process

The main equipment in this process is the strip mill. The following characteristics are recognised in each period of modernisation. The first period (until 1955) witnessed the birth of the hot strip mill. During this period three strip mills of a kind developed in the United States from 1926 to 1940 were in use, and older-style pull-over mills were also in operation. The annual production capacity was less than 200,000 tonnes and the speed of rolling was half that of present mills.

During the second period (1956–61), four strip mills of the type which had come to be standard in the United States from 1945 to 1960 were installed. Annual production capacity was raised to 1.5–2.0 million tonnes. In the third period, 1962–7, six very modern strip mills developed in the United States during the 1960s were installed. Enlargement of equipment, and high speed continuous operation, along with the introduction of AGC (Automatic Gauge Control) and computer control raised the annual production capacity to 3–4 million tonnes. In this period, the industrial machine industry also established its technological foundation. Home production of machine equipment, electric equipment and computer measuring equipment resulted in new improvements of imported technology for strip mills. During the fourth period, from 1968 onwards, seven new strip mills were installed. Enlargement of strip mills, high-speed operation, continuous operation and automated operation were further promoted (Tekkō Kyōkai 1976: 1).

Comparing the number of hot strip mills by country as of 1977, the United States had forty-six, Japan twenty-three, the USSR sixteen, West Germany eight, France six, the UK six and Italy six (Tekkō Renmei 1978).

AUTOMATION IN PRODUCTION SYSTEMS

Automated systems (automation) constitute the highest stage of mechanisation and are established by the thoroughgoing use of computers. In a large steel enterprise, automation is a comprehensive information processing system which has four major goals:

1 central control and timely processing of information using multipurpose computers;
2 a system of on-line real-time computers managing production (allowing high standards of process control and a large reduction of workforce);
3 automation of the main production process using computers;
4 integration of the computer processing into a system controlling total production and meeting the various orders from users.[6]

Such a comprehensive information process system based upon extensive use of computers raised the labour productivity of each production process and technically made it possible to produce new products. But as long as automation is inspired by profit-seeking and severe competition, the area where computers are used and the profitability of computer-usage varies. This creates an imbalance in the productivity of computerised production processes and non-computerised processes. As a result it has become more profitable to sub-contract such uncomputerised incidental and subsidiary work processes.

Development of computerisation

Computerisation in the iron and steel industry began in 1959 with the introduction of a Burroughs 101 computer by Yawata Steel Corporation. The number of installations of business computers (data processing computers) quickly rose between 1961 and 1963, then at a slower pace until 1966, when again computerisation proceeded apace until 1973 saw a downturn. This decline was due to the availability of larger and greater capacity computers.

The first process computer was introduced into the blooming process at the Chiba works of Kawasaki Steel Corporation in 1962. In the next year, processing computers were introduced into the sintering process at the Tobata works of Yawata Steel Corporation, the hot rolling process at the Sakai works of the same corporation, the LD converter process at the Kawasaki works of Nippon Kōkan KK, and the LD converter process at the Hirohata works of Fuji Steel Corporation; in 1964 into the blast furnace process at the Mizue works of Nippon Kōkan KK; and in 1965 into the open hearth process at the Kawasaki works of Nippon Kōkan KK and into the plate making process at the Tsurumi works of the same corporation (Tekkō Renmei 1969: 628). From 1968 the number of computers installed increased remarkably and the level of automation in production processes reached a high level in a short time.

The relationship between computerisation and the establishment of new integrated works along with the growth of steel output is shown over three periods. First, 1960–5 was a period when the punch card system and early computers coexisted. In this period the emphasis was on rationalisation of office work based upon machine processing of various statistics on order, production, shipment, etc. Second, 1965–70 is characterised by the development of various techniques of information processing, including the on-line system. And the third period, 1970–5, is the period when the development of an integrated management system to fit production and process in a new steel works was achieved (Tekkō Renmei, *Tekkō Kaihō* 1978: No. 1151). The establishment of new integrated works from 1965 onwards and the development of comprehensive information processing systems correspond in time. The output ratio of steel from these new works is increasing. Computerisation of newer works is more advanced; the degree of automation in the production processes, in quality control and in operation of equipment is greater; and the efficiency of such works is higher than in older ones (Tekkō Renmei, *Tekkō no IE*, May 1972: 29–54; Sumitomo Metal Industries 1977a: 296–305; Nippon Kōkan 1972: 577–84; Kawasaki Steel 1976: 488–99).

Areas and scope of automation

Modernisation of plant and equipment based upon computerisation has been positively promoted at the new integrated works of the major steel corporations. The nine major companies with blast furnaces account for 61 per cent of the business computers used in the steel industry. Six of them accounted for 92.8 per cent of the total process computers in January 1978 (Tekkō Renmei, *Tōkei Yōran* 1978). Thus automation of production processes is highest in the larger enterprises with blast furnaces. The use of business and process computers is examined in detail here.

The use of business computers in various areas is shown by the percentage of computer-use hours in Table 5.3. Process control accounts for 34.3 per cent, production and operation results account for 13.0 per cent, orders and sales for 9.5 per cent, shipping and transport 8.5 per cent, and in other areas where business computers are used the ratio is quite low.

Table 5.3 Working ratio of business computers by utilisation

Fields of utilisation	*Percentage of working hours of computers*
Demand forecast	0.0
Planning of business management	0.4
Planning of production	3.4
Process control	34.3
Production, operation results	13.0
Shipping, transport	8.5
Quality, technology	2.0
Orders, sales	9.5
Raw materials	1.2
Materials	2.1
Financial affairs, accounting	2.4
Cost accounting	2.5
Personnel, salaries	2.2
Estimate of plans	1.2
Estimate of technology	2.1
Managerial techniques	0.8
Information reference	1.4
Operation management	3.5
Others	9.5
Total	100.0

Source: Tekkō Renmei, *Tōkei Yōran* 1978: 224
Note: The above figures include fifty member companies of the Japan Iron and Steel Federation (as of January 1978)

The number of process computers and percentage of the total number for each process and plant is shown in Table 5.4. The process where the largest number of computers is used is the rolling process, accounting for 45.4 per cent of the total number of computers. This is followed by ironmaking, blooming and slabbing, raw material and steelmaking processes. The plant which has the largest number of computers and the highest percentage is the blast furnace, accounting for 16.2 per cent. This is followed by hot rolling, blooming and slabbing, bar steel, steel plate, LD converter and cold rolling plants. It is therefore the blast furnaces and hot rolling plants where the number of computers is highest.

Table 5.4 Process computers by processes, factory and ratio

Process	*Factory*	*Number of Units*	*Ratio %*
Raw material	Raw material	15	3.4
	Pellet	8	1.8
	Sintering	29	6.5
	Coke	6	1.4
Pig iron	Blast furnace	72	16.2
Crude steel	LD converter	34	7.7
Blooms, Billets,	Ingot making	49	11.1
slabs	Continuous casting	13	2.1
Rolling	Hot strip	59	13.3
	Cold strip	33	7.5
	Heavy plate	41	9.3
	Coating	8	1.8
	Rails, shapes, bars	47	10.6
	Pipes and tubes	13	2.9
Other	Energy centre	16	3.6
Total		443	100.0

Source: Tekkō Renmei, *Tōkei Yōran* 1977: 223
Note: The above figures include seven integrated iron and steel corporations, as of January 1977

The analysis of the use of computers at Nippon Steel Corporation works (Table 5.5) makes us further aware of the number of computers in new large-scale works such as Nagoya, Kimitsu and Oita, where the steel production capacity is also quite high. In these works a comprehensive information processing system has been established. Also, older large-scale works such as Yawata, Muroran and Hirohata, where steel production capacity is large compared with others of similar age, have more computers than smaller ones. The breakdown of computer use in terms of processes in Nippon Steel Corporation conforms with the general trend, in that the highest percentage of computer installation is the rolling process, accounting for 47.2 per cent. This is followed by ironmaking, raw material, blooming and slabbing, and steelmaking processes. The only difference found is that the raw material and blooming and slabbing processes are in reverse order. Comparing plants, the plant which has the highest rate of computer installation is the blast furnace plant and it accounts for 15.9 per cent. This is followed by hot rolling, bar steel, cold rolling, LD converter, steel plate and the blooming and slabbing plants.

Comparison of old and new works

The 1960s, particularly from 1965 on, was a time when major corporations competed in building new works, while at the same time modernising old works. When we compare new and old steel sites such as Kimitsu and Yawata, we find differences such as those given below. The corporate strategy was to narrow these differences through the modernisation of the old works.

To compare these locations, Kimitsu works is on a new sea-side plant nearest to the biggest market, while Yawata is away from demand markets and therefore packed distribution is dominant. Kimitsu has a very rational plant layout in terms of the flow of process (materials → production → shipping) as well as simplification of each production line, but Yawata has an irrational production process due to its combination of two separate sites. Kimitsu has the most modern, large, continuous, high-speed automated equipment and the actual amount of equipment is small, while Yawata has old, low-capacity equipment with many production lines. A comprehensive automation system extending from order entry to shipment exists at Kimitsu, while Yawata has no all-on-line system and process computers are only partially used. Kimitsu produces many products but main products are produced on a mass production system, while Yawata produces many products on small production systems and aims to specialise production towards high-grade steels with greater added value. Kimitsu has a large-scale sub-contract system with a smaller number of sub-contract firms, while Yawata has a small-scale sub-contract system with a large number of sub-contract firms. Kimitsu has difficulty in securing labour in its own district, and so is dependent upon the Kyushu labour market, while Yawata has the advantage of being able to secure labour in its own district.

Table 5.5 Number of computers in Nippon Steel Corporation, by steel works

Steel works	*Construction date*	*No. of employees*[a]	*Steel capacity – million /year*[b]	*No. of business computers*[c]	*No. of process computers*[c]	*Total no. of computers*[c]
Kamaishi	1874	3,924	1.8	2	8	10
Yawata	1901	20,142	8.56	4	26	30
Muroran	1909	6,501	4.6	4	23	27
Hirohata	1939	9,191	4.68	7	23	30
Hikari	1955	3,106	0.17	1	3	4
Nagoya	1958	8,630	7.2	3	36	39
Sakai	1961	3,512	4.5	3	17	20
Kimitsu	1965	7,669	10.0	9	39	48
Ohita	1971	3,699	8.0	3	30	33
Total		66,374	49.51	36	205	241

Sources: Nippon Steel, *Yukashōken Hōkokusho* 1977; Nippon Steel: Interview

Notes: (a) Number of employees, as of March 1977

(b) Steel capacity, as of 1976

(c) Number of computers, as of May 1977

Nippon Steel Corporation held back modernisation investment in the Yawata works from 1965 to 1970 in order to build the Kimitsu site. Thus, the important differences in terms of modernisation of these two works are that Kimitsu is greatly advanced in modernisation in terms of location, plant layout, equipment and automation; that differences in these four factors determine differences in production systems and products; and that these four factors, as well as the labour market, influence the skill level and scope of the sub-contract system.

These differences in the modernisation of the two representative works examined are broadly true of other steelworks as well. Namely, though there exist some minor differences due to the size of the works, new works such as Nagoya, Sakai, and Oita have a mass production system with a limited variety of products, while old works such as Kamaishi, Muroran and Hikari have a small production system with a large range of products, with an emphasis on high-grade steel items. This appears to be a characteristic of the modernisation of old and new steel works.

THE SUB-CONTRACT SYSTEM

Sub-contracting as a unique social division of labour in the Japanese steel works developed due to the difficulty of securing temporary workers during works modernisations. As already described, the third rationalisation plan emphasised the establishment of a number of new large-scale works and this required a bigger labour force. The means of securing an adequate labour force changed after 1960 from obtaining temporary labour to securing sub-contract labour (Tekkō Rōren 1971: 480; Itozono 1978: 116). The percentage of temporary labour of the total labour force employed by the fifty iron and steel enterprises declined remarkably from 13.0 per cent in 1960 to 0.5 per cent in 1976 (Tekkō Rōren, *Rōdō Handbook* 1977). The percentage of sub-contract labour greatly increased during the same period, as is shown in the case of the Muroran works of Nippon Steel Corporation, from 20.2 per cent in 1955 to 52.6 per cent in 1976 (Michimata 1978: 209). The displacement of temporary labour by the sub-contract system was also pursued in other places. The average rate of sub-contracting in a survey of forty-nine enterprises of the iron and steel industry accounted for 43.8 per cent of the total labour force as of December 1976. The average percentage of sub-contracting in new works (Kimitsu, Oita, Fukuyama, Kashima, Kakogawa and Mizushima) of the five major corporations accounted for 57.1 per cent, i.e. more than half of the labour force (Tekkō Rōren, *Rōdō Handbook* 1977).

Development of the sub-contract system

Factors which worked for the growth of the sub-contract system with regard to modernisation of plant and equipment are that iron and steel enterprises needed large numbers of inexpensive but adaptable workers and also the propensity to adjust their workforces according to changes in the economic situation; and that many firms with small capital which had previously undertaken sub-contract work at the plants were able to develop their experience and technique in special fields, both in engineering sub-contracting and regular work sub-contracting, and succeeded in enlarging their capital by mergers and reorganisations among themselves as well as by the increased contracts created by the modernisation and construction projects of parent companies.

Three reasons can be given to explain the higher proportion of contracting in the new works, where the level of modernisation and automation is higher than in old ones:

1 as the modernisation of plant and equipment raised the amount of fixed capital rapidly to very high levels, the enterprises were urged to reduce labour costs as much as possible in order to minimise total capital expenditure;
2 it was possible to expand the area of sub-contracting in new works without industrial relations problems, because experienced workers from old works could be carefully chosen and new workers employed;
3 the high level of automation and the expansion of areas where computers are used lowered the need for skilled labour and simplified the complexity of many specific jobs, thus creating areas of work available for sub-contracting.

In addition, there were several advantages in the use of sub-contracting, such as: the workforce can be adjusted according to the volume of work; the risk of labour disputes, as sub-contracted tasks are generally manual and unskilled and wages in such firms are usually low, can be passed on to sub-contracted firms; the differences that exist in job content, wage level and status engender an elite and loyal consciousness among workers in steel corporations; and sub-contracted labour is not eligible for welfare care or retirement lump-sum bonuses given to the corporations' regular workers (Ishida 1967: 104). Among the methods of reducing the workforce of Nippon Steel Corporation from 1966 to 1973, sub-contracting was more effective than other methods, such as mechanisation and reallocation.

Sub-contract system by plant

Table 5.6 shows the average percentage of sub-contracting in the different plants and departments of twenty works. The average percentage of sub-contracting is 52 per cent; it means that more than half of the jobs in the works are done by sub-contract workers. The average percentage of sub-contracting of various plant production jobs (direct department sub-contracting) is 41 per cent, while sub-contracting of peripheral services (indirect department sub-contracting) is 64 per cent. Thus, sub-contracting is more advanced in indirect department sub-contracting, where work is more labour-intensive. In indirect department sub-contracting, the analysis and experimental section where

Table 5.6 Sub-contract rate of twenty steel works by department (1975)

	Employees (A)	*Sub-contract workers (B)*	*Total (C)*	*Sub-contract rate (B/C), %*
Sintering line	3,671	3,217	6,888	47
Blast furnace	3,552	2,698	6,250	43
Converter	9,101	6,127	1,528	40
Continuous casting	2,596	1,598	4,194	38
Blooming, slabbing	6,905	8,784	15,689	56
Plate production	5,635	3,015	8,650	35
Hot strip	6,280	3,174	9,454	34
Cold strip	8,988	4,064	13,052	31
Coating	2,377	1,210	3,587	34
Large shapes	3,120	1,736	4,856	36
Coil	2,078	1,653	3,731	44
Steel bar	1,486	1,508	2,994	50
Tube	4,247	2,547	6,794	37
Total of direct departments	60,036	41,331	101,367	41
Receipts and payments for materials	1,196	6,491	7,687	84
Receipts and payments for	1,887	9,917	11,801	84
manufactured goods	4,053	222	4,275	5
Analysis and experiment	18,368	27,258	45,626	60
Repair	4,699	1,046	5,745	18
Power	3,545	14,867	18,412	81
Transport				
Total of indirect departments	33,748	59,801	93,546	64
Grand total	93,784	101,132	194,913	52

Source: Yamashita 1977

expertise of a high educational level and responsibility are required, and the power section, which has a very important role for the whole works, have very low percentages of sub-contracting. By contrast, the sections for receiving and shipping of raw materials and manufactured goods, repair and transport have a very high sub-contracting percentage. In direct department sub-contracting, the percentage of the various production jobs sub-contracted is rather similar, except for the slabbing and blooming and steel bar plants, where it is over 50 per cent.

The difference in the sub-contracting percentages of the various production jobs in direct departments is caused by differences in the level of modernisation of each plant, differences in the amount of auxiliary work incidental to the work of the various main processes, and also the quantity of labour-intensive work. Auxiliary work in plants where the sub-contracting rate is high is usually labour-intensive and requires skilled labour. For example, in over nineteen hot rolling plants surveyed, almost all the work of the hot rolling line is done by workers employed by the steel firms, while much of the work in the roll shop and on the finishing line is done by sub-contract workers. In the roll shop and on the finishing line, operation, inspection and management are done by steel firm workers, while various technical jobs which require technical qualifications, such as welding and packing work, are done by sub-contract workers.

The sub-contract system at Kimitsu works

Kimitsu, which is representative of the new steelworks, had a sub-contracting rate of 64.1 per cent in December 1976. Its sub-contract systems are examined here, paying attention to their content and relative importance in the works.

This works was established by Yawata Steel Corporation in 1967 in order to deal with the expanding market in the Kanto area. Today its steel production capacity is 10 million tonnes per year. It is the biggest works of the Nippon Steel Corporation.

The corporation pursued the maximum level of sub-contracting from the start, making it a rule to sub-contract all incidental lines (Tekkō Renmei, *Tekkō no IE*, 1973: No.1: 35). The sub-contract system of this works is composed of many companies specialising in certain areas, as shown in Table 5.6. They are: subsidiary companies; companies doing sub-contracting work in the production process; sub-contracted companies doing transport work; companies maintaining equipment and supplies; sub-contracted companies doing special equipment and supplies; companies performing business services and maintenance; and others. Companies sub-contracted for production process work supply a great deal of the labour force, playing an important role in the works, as shown in Figure 5.5.

In the steel works the division of labour is established by choosing the most suitable sub-contract firms for the type and amount of work in each main process. The extent and depth of division of work is greatly advanced, so that sophisticated synchronisation of all functions and their control are required in order to keep the works running without inconvenience. A good information network with the use of computers is essential, as is some mechanism for maintaining good industrial relations, not only in the steel company but also in sub-contract firms. This is a prerequisite for any extensive work division based upon a number of sub-contract firms.

Table 5.7 shows the classification of these companies by type of work, amount of capital, number of employees, number of shares held by the Nippon Steel Corporation, and companies which expanded to Kimitsu from Yawata works. Among those sub-contract companies undertaking regular work incidental to each main process, companies supplying a great deal of the labour force are large firms with capital of more than ¥100 million. Transport companies such as Sankyu, which deals with raw materials, are large in both capital and workforce and the steel company does not hold any shares in these. Companies whose shares are held by the parent company are those which provide services and labour which are as essential as regular employees. It seems that the further the function of sub-contract companies from the core production technology, the more independence they have from Nippon Steel Corporation and their commitment to Kimitsu works becomes only a part of their whole business.

The above division of labour is shown in Table 5.8 in terms of the total workforce of the steelworks. The employees of Nippon Steel Corporation account only for 35.9 per cent, while the sub-contract firms account for 64.1 per cent. In terms of production workers, corporation staff account for 26.4 per cent, while the sub-contracted firms account for 57.4 per cent. The percentage of clerical and technical personnel of Nippon Steel Corporation is only 9.5 per cent, in which managerial employees are also included. On the other hand the percentage of sub-contract workers for engineering work is as high as 6.7 per cent, suggesting that the majority of less important engineering work is now sub-contracted. This implies that university graduate technical employees of Nippon Steel Corporation are mostly fulfilling the managerial function of supervision and control of the engineering activities carried out by the sub-contractors. This dual structure of workforce

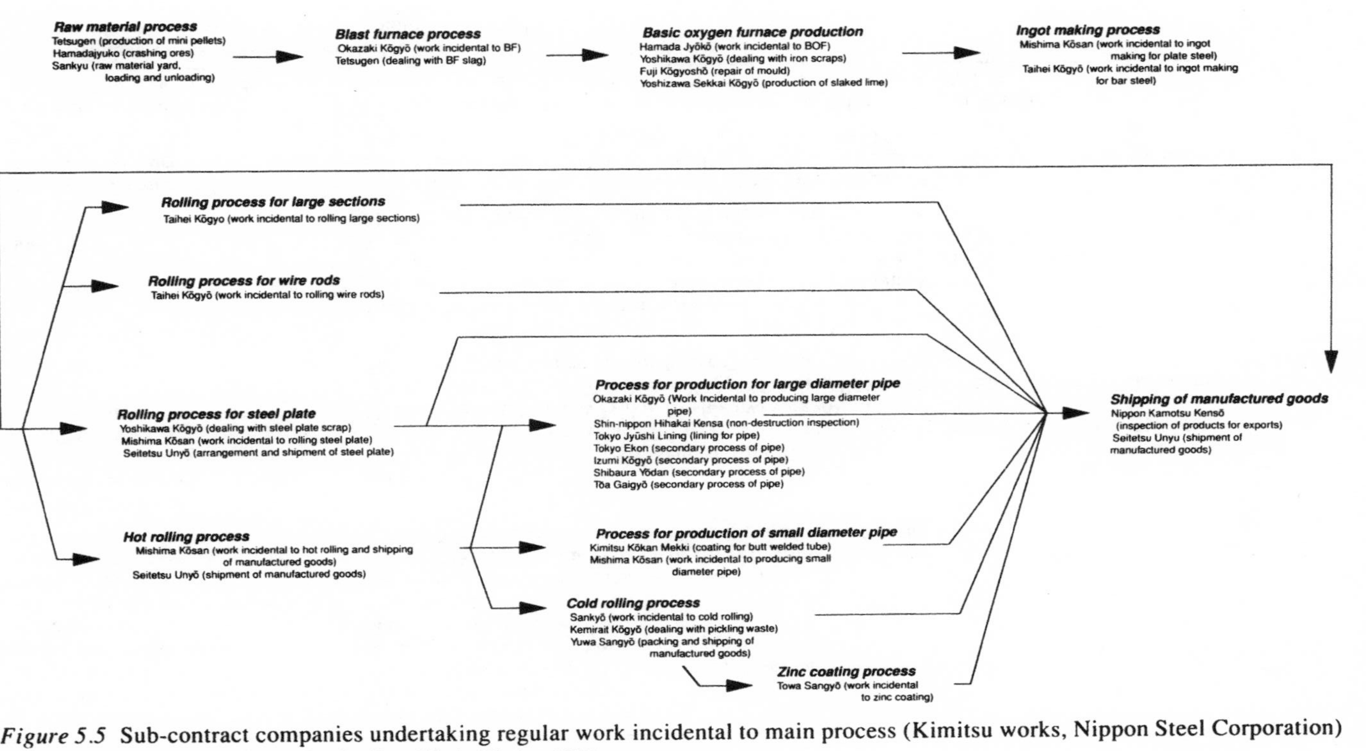

Figure 5.5 Sub-contract companies undertaking regular work incidental to main process (Kimitsu works, Nippon Steel Corporation)
Source: Kimitsu works (Nippon Steel), *New Life in Kimitsu* 1974

Table 5.7 Classification of sub-contracting firms undertaking regular work

1) Type of Work	*Name of Company*
Work related to raw material,	Tetsugen, Yoshizawa Kōgyō
subsidiary raw material	Sankyu, Okazaki Kōgyō, Taihei Kōgyō, Mishima
Regular work incidental to	Kōsan, Hamada Jyūkō, Yoshikawa Kōgyō
main process	Seitetsu Unyū
Work related to transportation	Yuwa Sangyō
Packing	Tokyo Ekon, Tōa Gaigyō, Izumi Kōgyō, Kimitsu
Secondary process work	Kōkan Mekki, Shibaura Yōdan Kōgyō, Fuji Kōgyōsho, Kemirait Kōgyō, Tokyo Jyūshi
Others	Lining, Nippon Kamotsu Kensū, Shin nippon Hihakai Kensa
2) Amount of Capital	
over 5,000 million yen	Sankyū
1,000–5,000 million yen	Okazaki Kōgyō, Taihei Kōgyō, Seitetsu Unyu Tetsugeñ,
100–1,000 million yen	Kemirait Kōgyō, Tokyo Ekon, Tokyo Jyūshi Lining, Yuwa Sangyō, Yoshizawa Sekkai Kōgyō, Mishima Kōsan, Hamada Jyukō, Yoshikawa Kōgyō
under 100 million yen	Tōa Gaigyō, Fuji Kōgyōsho, Izumi Kōgyō, Kimitsu Kōkan Mekki, Shin nippon Hihakai Kensa, Shibaura Yōdan, Nippon Kamotsū Kensū Kyōkai
3) Number of Employees	
Over 1,000	Sankyū, Seitetsu Unyū, Mishim a Kōsan
500–1,000	Okazaki Kōgyō, Taihei Kōgyō, Yuwa Sangyō, Hamada Jyūkō
100–500	Tetsugen, Tokyo Jyūshi Lining, Yoshikawa Kōgyō, Fuji Kōgyōsho, Kimitsu Kōkan Mekki
Under 100	Kemirait Kōgyō, Tokyo Ekon, Yoshizawa Sekkai Kōgyō, Tōa Gaigyō, Izumi Kōgyō, Shin nippon Hihakai Kensa, Nippon Kamotsu Kensa Kyōkai.
Percentage of shares held by Nippon Steel Corporation	Yuwa Sangyō (97.6), Kemirait Kōgyō (52.5), Taihei Kōgyō (43.5), Tokyo Jyūshi Lining (31.5), Seitetsu Unyū (30.0), Tokyo Ekon (30.0), Tetsugen (28.0), Kimitsu Kōkan Mekki (20.0), Tōa Gaigyō (10.0), Okazaki Kōgyō (5.2)
Companies which expanded to Kimitsu from Yawata works	Sankyū, Okazaki Kōgyō, Tetsugen, Yuwa Sangyō, Mishima Kōsan, Hamada Jyūko, Yoshikawa Kōgyō, Fuji Kōgyōsho, Nippon Kamotsu Kensu Kyōkai

Source: Kimitsu works (Nippon Steel) 1974; Tokyo Shōkō Research 1978
Note: Amount of capital and number of employees, as of 1974

Table 5.8 Workforce composition of Kimitsu works (Nippon Steel Corporation)

Workforce at Kimitsu works	*Number of personnel*	*Percentage*
Nippon Steel clerical and technical personnel	2,058 (466)	9.5
Nippon Steel regular workers	5,738 (17)	26.4
Sub-total	7,796 (483)	35.9
Sub-contract firm workers for engineering	1,467 [17]	6.7
Sub-contract firm workers for regular work	12,473 [42]	57.4
Sub-total	13,940	64.1
Total	21,736	100.0

Source: Tekkō Rōren, *Rōdō Handbook* 1977: 10–11, 30–1
Notes: (a) Number of employees of Nippon Steel Corporation, as of April 1977 and the number of sub-contract firms and workers, as of December 1976
(b) ( ) the number of females included in the above figures, [ ] the number of sub-contract firms

forms a framework of division of labour in the new steel works. It also provides a framework for management and industrial relations.

CONCLUSION

The major companies of the Japanese steel industry took great advantage of surrounding global economic conditions in their ambitious and determined modernisation programmes. In addition to modernising older steelworks, the various corporations also built new production sites and developed a mode of production utilising high-speed and continuous production at high levels of capacity. To ensure the labour flexibility required by this new technology, a unique 'dual workforce' system, distinguishing between regular employees with full company benefits and sub-contracted workers for whose welfare the corporations were not responsible, was devised and extensively put into practice.

This whole process, which continued until the years 1971–5 (see Table 5.1), is another illustration of the process of organic compatibility.

6 Management and labour in Japanese steel works

The new production system, with its comprehensive on-line information technology, ushered in various changes in the make-up of management and labour at works level. The idea of workplace management was originally introduced from the United States in the 1950s as part of a productivity movement in Japan[1] and during the modernisation process was modified and developed into the system we see today. Workplace management under the new production system is an integral part of the whole corporate management system, which is under head office control. Much of the analysis which follows is relevant to the proposed conceptual framework of relative 'advance' and 'convergence'.

This chapter will make use of an extensive analysis of changes in workforce composition to assess these developments at the steel works.

CHANGES IN WORKFORCE COMPOSITION AND MANAGEMENT

Modernisation at a steel works involves factors of size, speed, continuity of production processes, and the extent and level of automation in the works. Embarking on such a programme allows management the simultaneous opportunity to manipulate workforce composition. Table 6.1 compares the workforce composition of a representative old works (Yawata) in 1960 and a new works (Kimitsu) in 1975. The percentage of managerial personnel above section chief rose from 0.3 to 1.6 per cent and the percentage of technical and clerical personnel rose from 16.3 to 24.1 per cent, while the percentage of blue-collar workers declined from 83.4 to 74.3 per cent.

We can assume that administrative functions shifted as a whole from blue-collar to white-collar employees, and the relative importance of technical personnel within administrative divisions was enhanced.[2]

Such employees are those with a technological background from university, and they are responsible for the overall operation of the works as well as for troubleshooting in both operational areas and major equipment.

Table 6.1 Changes in workforce composition of steel works

	1960 (Yawata works)		*1975 (Kimitsu works)*	
	Number	*Ratio %*	*Number*	*Ratio %*
Managerial personnel	132	0.3	124	1.6
Clerical and technical personnel	6,144	16.3	1,895	24.1
Workers	31,470	83.4	5,840	74.3
Total personnel	37,746	100.0	7,859	100.0

Source: Yawata Steel Corporation 1961; Kimitsu works (Nippon Steel) 1975; Tekkō Shimbun-sha 1960; Tekkō Rōren, *Rōdō Handbook* 1975

Note: (a) Number of personnel of Yawata works, as of June 1960. Number of administrative personnel of Kimitsu works, as of February 1975, number of clerical and technical personnel, workers and total personnel of the same works, as of June 1975

(b) Administrative personnel includes section chiefs and all higher administrative staff

The drastic decrease in the total workforce is due both to the modernisation of production processes, represented by the new works – and in particular its continuous production processes – and an increase in sub-contracting in the extended areas peripheral to the main processes. Administrative functions have become more specialised and this expansion and specialisation resulted in an increase of managerial personnel. In addition, the ratio of managerial to technical and clerical personnel calculated from Table 6.1 increased from 2.1 per cent in 1960 to 6.5 per cent in 1975; thus managerial staff increased relatively more than technical and clerical personnel.

If we examine the increased percentage of managerial personnel of the same works in connection with changes in organisation, we find three major changes in terms of the structure of the organisation.

First, there were eighteen organisations (thirteen departments and five other organisations) in 1960, while in 1975 there were fourteen (eleven departments and three other organisations). Thus we can find a certain rationalisation of organisation in modernised new works. The functions of the management bureau, which existed in 1960, had been assumed by various indirect departments (system development, production service, equipment, technology) and others by 1975. Also by 1975, the functions of the labour and education departments of 1960 had been combined as the organisation and personnel department; the 1960 accounting and business department's functions were to be found in the general affairs department, transport in the production service department, inspection and technology research department functions in the technology department, power, civil engineering, engineering and equipment department functions in the equipment department.

Second, organisations which appeared between 1960 and 1975 are the section for environment management, and the system development department. These organisations were established to deal with public hazards caused by equipment enlargement and 24-hour operation, and the necessity for information processing systems following on from automation in the production and development of on-line systems.

Finally, by 1975, independent rolling departments (bar steel, hot rolling, cold rolling, and steel pipe departments) had been created, reflecting the expansion and division of rolling processes caused by the mass production of iron and steel, and also the need to meet the diversification of orders for steel products.

In terms of personnel composition, the organisation with the highest number of personnel in 1960 was the management bureau (15.2 per cent), followed by the construction, technology research, transport and iron production departments. The organisation with the highest number of personnel in 1975 was the technology deptartment (23.4 per cent), followed by equipment, general affairs, production service, organisation and personnel and steel pipe departments. The combined personnel in the technology, equipment and production service departments account for almost 50 per cent, showing the effect of large-scale plant and equipment modernisation, which requires many employees for the service and care of modern plant and equipment. The number of personnel in the general affairs department is high, for it includes accounting, financial affairs, and supply sections in addition to the general affairs, secretarial, publicity and public relations sections, thus showing this department's increase in importance. The organisation and personnel department includes organisation, personnel, labour, welfare, and safety sections. The organisation section includes an office for managing sub-contracting affairs and a personnel section, including an office for developing employees' ability (the faculty development office), which shows an integration of the labour department and education department functions which existed in 1960. The inclusion of offices for managing sub-contracting affairs and for developing the ability of employees respectively reflect the expansion of sub-contracting during plant and equipment modernisation and the characteristics of labour management in the 1970s. The high number of personnel in the steel pipe department in 1975 indicates that the rolling organisation had expanded and diversified, thus increasing their relative importance, while ironmaking and steelmaking departments had been rationalised. Of the rolling departments, the importance of the steel pipe department increased in terms of the number of personnel.

The organisational characteristics of the new works can be summarised in the following points:

1 production-related departments (iron production, steel production, bar steel, hot rolling, cold rolling, and steel pipe) increased, accounting for 27.3 per cent of total personnel;
2 the technology and equipment departments, which were set up to maintain, repair and improve the modernised equipment, account for as much as 41.2 per cent of the total staffing;[3]
3 a system development department had come into being, reflecting the high level of automation;
4 division of labour between works and head office developed and a significant part of the managerial functions of the works had been absorbed into head office by the use of on-line systems.

These characteristics reflect the impact of modernisation and the nature of the management response.[4]

In Table 6.2 we compare managerial hierarchies in both works and consider the implications of changes. The control span of management, as calculated from Tables 6.1 and 6.2, was 286:1 in 1960, while it was 63:1 in 1975, providing conditions for tighter and more sophisticated labour control. The number of managerial positions was fourteen in 1960, decreasing to nine in 1975, which resulted in a flat line management structure. The decrease in the number of positions is greatest in senior management, reflecting the rationalisation in managerial hierarchy. As to changes in the number of people in the various positions, in 1960 managers accounted for 12.0 per cent of total managerial personnel, and the senior management category stood at 18.2 per cent; both had declined, to 8.9 and 13.7 per cent respectively, by 1975.[5] This shift increased the span of control of senior management from 4.5 to 6.2 per cent and implies a slight increase in the area and amount of their responsibilities. In addition, the decline in the percentage of senior management positions suggests that promotion to these line positions had become more difficult by 1975.[6] Consequently, some way was needed to avert decline in morale among staff who failed to secure promotion, and the *Nōryokushugi Kanri* system was one response

Table 6.2 Composition of managerial personnel in large-scale steel works, by managerial position

Managerial position	*1960 (Yawata works) No. of personnel*	*Ratio (%)*	*Managerial position*	*1975 (Kimitsu works) No. of personnel*	*Ratio (%)*
Top management			*Top management*		
Head of steel works	1	0.8	Head of steel works	1	0.8
Sub-total	1	0.8		1	0.8
Senior management			*Senior management*		
Manager	16	12.0	Manager	11	8.9
Assistant head of steel works	2	1.5	Assistant head of steel works	2	1.6
Head of engineers	1	0.8	Room chief	4	3.2
Room chief for head of steel works	1	0.8			
Head of research institute	1	0.8			
Head of bureau	2	1.5			
Head of hospital	1	0.8			
Sub-total	24	18.2		17	13.7
Assistants to senior management			*Assistants to senior management*		
Assistant chief	7	5.3	Assistant manager	10	8.1
Assistant manager	3	2.3	Assistant specialist manager	3	2.4
Assistant head of hospital	2	1.5			
Sub-total	12	9.1		13	10.5
Middle management			*Middle management*		
Section chief	82	62.1	Section chief	80	64.5
Chief of research room	12	9.0	Head of factory	12	9.7
			Head of clinic	1	0.8
Chief of branch office	1	0.8			
Sub-total	95	71.9		93	75.0
Total	132	100.0		124	100.0

Sources: Tekkō Shimbun-sha 1960; Kimitsu works (Nippon Steel) 1975
Note: Administrative personnel includes section chief and all higher administrative staff

adopted in the 1970s. Nonetheless, this did not fully alleviate the promotion competition among section chiefs.

In general, the formal management structure became simpler, but with the introduction of *Nōryokushugi Kanri* the control of managerial staff itself became more sophisticated, resulting in an increased hierarchy of finely-graded promotions. In particular, as elevation under *Nōryokushugi Kanri* involves a larger element of assessment by one's immediate superiors, those in lower management have become more concerned about their personal relationships at work, in particular with the line managers above them. At the same time, their tasks and responsibilities as section chiefs have increased.

To identify a more precise relationship between plant and equipment modernisation and workforce composition, Table 6.3 compares the total workforce composition of an old large-scale steelworks (Yawata) with a new one (Kimitsu) in 1977. Although almost the same in terms of annual steel output (6.88 million tonnes for Yawata and 6.76 million tonnes for Kimitsu), the total number of employees is very different, at 41,235 and 21,736 respectively. Labour productivity at Kimitsu is therefore nearly twice as high as at Yawata when measured against crude steel output in 1975.

The degree of sub-contracting is influenced by the level of works modernisation, being higher, at 64.1 per cent, in the newer example. This higher percentage of sub-contracting at Kimitsu makes the percentage of technical and clerical personnel 2.5 per cent lower than at

Table 6.3 Comparison of workforce composition of old and new large-scale steel works (Nippon Steel Corporation)

Classification	*Old large-scale steel works (Yawata) 1977*		*New large-scale steel works (Kimitsu) 1977*	
	No. of personnel	*Ratio (%)*	*No. of personnel*	*Ratio (%)*
Employees of Nippon Steel				
Clerical and technical personnel	4,972 (786)	12.0	2,058 (466)	9.5
Workers	17,969	43.6	5,738	26.4
Sub-total	22,941 (485)	55.6	7,796 (17)	35.9
Sub-contract workers				
Sub-contract workers for regular work	16,073 [128]	39.0	12,473 [42]	57.4
Sub-contract workers for engineering	2,221 [16]	5.4	1,467 [17]	6.7
Sub-total	18,294	44.4	13,940	64.1
Total	41,235	100.0	21,736	100.0

Source: Tekkō Rōren, *Rōdō Handbook* 1977

Notes: (a) Number of employees of Nippon Steel Corporation, as of April 1977 and number of sub-contract firms and workers, as of December 1976

(b) ( )(the number of females included in the above figure, [] the number of sub-contract firms

Yawata, and the proportion of regular workers is 17.2 per cent less than in the older site. Conversely, the new large works has 18.4 per cent more sub-contract workers undertaking regular work and 1.3 per cent more undertaking engineering work than the old large works. The number of sub-contract firms for engineering work is almost the same in both, but the number of firms doing regular work is different, with 128 firms at Yawata and forty-two at Kimitsu; however, the number of workers per sub-contracting firm is 125 for the former and 297 for the latter, showing greater size of sub-contract firms in the new large works. These companies grew as the sub-contracted areas expanded in the new steel works. Fairly big sub-contract firms, which could afford modern machinery and equipment for their own operations, were carefully chosen with a view to the necessity of technological integrity at the works (Tekkō Renmei, *Tekkō no IE* 1973: 35).

The larger percentage of sub-contracting in the new works can be explained both by the difficulty of recruiting regular employees in Kimitsu itself, which thus favoured the replacement of regular jobs by sub-contract workers, and the weakness of the unions at the site, which created favourable conditions for management to expand extra-company employment. At Yawata, any expansion of sub-contracting was rather more problematic, as it could only be done by sacrificing regular employees of the Nippon Steel Corporation. Thus the new works had, in effect, a 'green field' for the establishment of sub-contracting contracts.[7]

This process of expansion in sub-contracting encouraged sub-contracting firms themselves to modernise their plant and equipment, thereby affecting the content of work and skills of their workers. Nonetheless, as a whole their jobs are relatively more labour-intensive and physically harder than those of regular employees. The larger percentage of sub-contracting however suggests that the role and importance of sub-contracting has increased in the new works and it is interesting to note that there the attitudes of sub-contract workers are more positive and confident than the ones in the old works. In Yawata the workers of Nippon Steel are more positive and self-confident than the sub-contract workers.

Sub-contracting has introduced another important factor into the works management equation. Nippon Steel managers have had their task of ensuring co-operative industrial relations eased in their own sphere by the dual workforce structure, in that it has transferred industrial relations management of more than half of its active production workers to sub-contract firms.

Tables 6.4 and 6.5 compare the workforce composition of the old and

Table 6.4 Workforce composition of old and new large-scale steel works by clerical and technical personnel and workers (1977)

	Old large-scale steel works (Yawata)		*New large-scale steel works (Kimitsu)*	
	No. of personnel	*Ratio (%)*	*No. of personnel*	*Ratio (%)*
Clerical and technical personnel	4,972	21.7	2,058	26.4
Workers	17,969	78.3	5,738	73.6
Total	22,941	100.0	7,796	100.0

Source: Data from Table 6.3
Note: Number of personnel, as of April 1977

Table 6.5 Composition of managerial personnel in old and new large-scale steel works by clerical, technical and production departments (1975)

	Old large-scale steel works (Yawata)		*New large-scale steel works (Kimitsu)*	
	No. of personnel	*Ratio (%)*	*No. of personnel*	*Ratio (%)*
Clerical department	64	26.5	29	23.4
Technical department	98	40.4	61	49.2
Production department	81	33.5	34	27.4
Total	243	100.0	124	100.0

Sources: Tekkō Shimbun-sha 1975; Kimitsu works, Nippon Steel 1975
Notes: (a) Managerial personnel includes section chiefs and all higher managerial employees
(b) Number of personnel in old large-scale steel works, as of May 1975; that of new large-scale steel works, as of February 1975

the new works, excluding sub-contract workers. The percentage of technical and clerical personnel at the newer site is 4.7 per cent higher than the old. Managerial personnel (above section chiefs) composition by clerical, technical and production departments shows that the managerial element in clerical areas is 2.9 per cent lower in the new works than in the old works; while for the technical department it is 8.8 per cent higher and for the production department 5.9 per cent lower in the new works. Kimitsu shows a high percentage of managerial personnel in the technical department, at just under 50 per cent, reflecting the expansion and importance of managerial functions in the technical department. We can identify the upgraded importance of the technical department as an impact of the new production system upon management and labour. The decline in the percentage of managerial personnel in the clerical department reflects the rationalisation of clerical managerial functions in the works through the use of on-line computers connected with the central computer at the head office. The decline in the percentage of managerial personnel in the production department also reflects the rationalisation of the workforce due to plant and equipment modernisation in the main processes, mainly through the integration and continuity of production processes. In the old works the number of managerial personnel is large in the production department, due to the large number of small plants in the works. In particular, rolling plants are small but numerous in the old works, and the other way round in the new works.

Thus modernisation shifted the importance of managerial employees from production and clerical to technical areas, whose functions are to maintain the large-scale, continuous, high-speed and automated production system and ensure its smooth operation.

Table 6.6 shows how such changes in the above composition of managerial personnel are reflected in the situation of labour management in terms of the control span of a manager. There is a significantly lower span at the new works, which suggests that tighter labour management exists there. Production operatives are more widely distributed and are now in a smaller span of management. The individual worker has more contact with management than with unions, which furthers the development of 'harmonious' industrial relations. Employees in the new works appear more individual and rational in thinking than those in the old works.[8]

The male–female workforce composition will be examined here in connection with the modernisation of plant and equipment. Table 6.7 shows that the total percentage of female employees is slightly higher in the new large-scale works. Among clerical and technical employees,

Table 6.6 Number of employees per single managerial personnel member in old and new large-scale steel works (1975)

	Old large-scale steel works (Yawata)	*New large-scale steel works (Kimitsu)*
Clerical and technical employees	30.4	22.4
Workers	249.8	171.8
Average of all employees	103.5	63.4

Source: Compiled from Table 6.2; Tekkō Rōren, *Rōdō Handbook* 1975
Note: Administrative personnel include section chiefs and all higher administrative staff

Table 6.7 Composition by sex of old and new large-scale steel works

	Old large-scale steel works (Yawata)			*New largescale steel works (Kimitsu)*		
	Male	*Female*	*Total*	*Male*	*Female*	*Total*
Clerical and technical						
Number of personnel	4,186	786	4,972	1,592	466	2,058
(Ratio %)	(84.2)	(15.8)	(100.0)	(77.4)	(22.6)	(100.0)
Workers						
Number of personnel	17,484	485	17,969	5,721	17	5,738
(Ratio %)	(97.3)	(2.7)	(100.0)	(99.7)	(0.3)	(100.0)
Total number of personnel	21,670	1,271	22,941	7,313	483	7,796
(Ratio %)	(94.5)	(5.5)	(100.0)	(93.8)	(6.2)	(100.0)

Source: Tekkō Rōren, *Rōdō Handbook* 1977
Note: Number of personnel, as of April 1977

it is 6.8 per cent higher in the new works, accounting for 22.6 per cent; while among production workers, it is 2.4 per cent lower, accounting for only 0.3 per cent. At the new works, owing to the automation of office work through the use of computers, more jobs have apparently been allocated to women in clerical and technical areas. These female employees are mostly high school or junior college graduates recruited from the local area, while the male employees in this category are mostly university graduates recruited from various areas. In the production area, due to the high level of automation and the continuous three-shift/four-group production system, the working area for women has been narrowed; this is partly attributable to the law which prohibits women from working night shifts.[9] Women are therefore only employed in areas such as the experimental and testing processes, cafeteria and clinic.

We have already examined in Chapter 5 the characteristics of placement of sub-contract firms in the new large-scale works, and we know that the use of sub-contracting is far greater in the new works. Here we investigate the workforce composition of representative sub-contract firms in the new large-scale works and examine its relation to plant modernisation.

The six sub-contract firms shown in Table 6.8 were all originally main sub-contract firms at the Yawata works, and at present they undertake regular incidental work at Kimitsu.[10] The amount of capital of these firms ranges from ¥0.1 to ¥3.7 billion. If we compare the percentage of clerical and technical personnel as a part of total personnel, the six sub-contract firms have a lower proportion than that of the new large-scale works. The percentage of production workers in the six sub-contract firms is higher than that of the new works. The percentage of technical personnel (Technical Department and Equipment Department) of the Kimitsu works is higher than that of sub-contract firms. In the case of the six sub-contract firms, although the firms are comparatively large and have quite considerable fixed capital investment in the works, most of their work takes place in non-automated areas, albeit partially and sometimes completely mechanised. There is no necessity for their having a technical personnel department like that of the new works itself. The work done by the sub-contract firms is mostly labour-intensive and the workers are mostly skilled manual labour, as examined in a previous chapter. Thus the division of labour in the new works is also reflected in workforce composition, suggesting that sub-contract workers are responsible for the labour-intensive peripheral labour.[11]

The percentage of women in both the new works and the six sub-

Table 6.8 Workforce composition of new large-scale steel works (Kimitsu) and total employees of its six sub-contract firms

	New large-scale steel works (Kimitsu)		*Total of six sub-contract firms*	
	No. of personnel	*Ratio (%)*	*No. of personnel*	*Ratio (%)*
Clerical and technical	2,058 (466)	26.4 (6.0)	625 (95)	15.3 (2.3)
Workers	5,738 (17)	73.6 (0.2)	3,445 (150)	84.7 (3.7)
Total	7,796	100.0	4,070	100.0

Source: Tekkō Rōren, *Rōdō Handbook* 1977, 1978

Notes: (a) Number of personnel of new large-scale steel works is as of April 1977, while that of total of six sub-contract firms is as of April 1978

(b) The six sub-contract firms are Okazaki Kōgyō, Hamada Jūkō, Taihei Kōgyō, Yuwa Sangyō, Mishima Kōsan and Yoshikawa Kōgyō

(c) ( ) the number of females included in the above figure

contract firms is rather small, accounting for about 6 per cent of the total (Table 6.9). This arises because the sub-contract firms are also operating the three-shift/four-team system. The amount of clerical work, especially general office work, is also rather small. The percentage of women in clerical and technical personnel is higher in the new works than in the six sub-contract firms, while in terms of production personnel, it is the other way around. This may be explained by the expansion of jobs caused by the introduction of office computers, considered to be an area of female employment, in the new large-scale works, while in the six sub-contract firms mechanisation and automation of office work is not advanced, and the percentage of women is lower. On the contrary, female participation in production is higher in the sub-contract firms, due to the labour-intensive work, and there exists some rather simple auxiliary work which women may undertake in the day shift.

The average age and length of service of employees at the new and old large-scale works are shown in Table 6.10. In 1978, in the new works, where modernisation of plant and equipment was more advanced, the average age was 7.6 years and the average length of service 7.1 years lower than in the old large-scale works. This was advantageous in saving labour costs, as wages are mainly determined by these factors. However, the fact that the average age in the old large-scale works was 40.6 and the average length of service 20.5 years suggests that seniority-based labour management could no longer be considered in the best

Table 6.9 Male and female workforce composition of new large-scale steel works and six representative sub-contract firms

	New large-scale steel works (Kimitsu)			*Total of six representative sub-contract firms*		
	Male	*Female*	*Total*	*Male*	*Female*	*Total*
Clerical and technical personnel	1,592	466	2,058	530	95	625
	(77.4)	(22.6)	(100.0)	(84.8)	(15.2)	(100.0)
Workers	5,721	17	5,738	3,295	150	3,445
	(99.7)	(0.3)	(100.0)	(95.6)	(4.4)	(100.0)
Total	7,313	483	7,796	3,825	245	4,070
	(93.8)	(6.2)	(100.0)	(93.4)	(6.0)	(100.0)

Source: Tekkō Rōren, *Rōdō Handbook* 1977, 1978

Notes: (a) Number of personnel of new large-scale steel works is as of April 1977

(b) The six sub-contract firms are Okazaki Kōgyō, Hamada Jūkō, Taihei Kōgyō, Yuwa Sangyō, Mishima Kosan and Yoshikawa Kōgyō. The number of employees is as of April 1978, except Yoshikawa Kōgyō which is as of April 1976

(c) () percentage

interests of rationalisation, and the average age and length of service increased to 45.3 and 25.7 respectively by 1992 (Tekkō Rōren, *Rōdō Handbook* 1992).

We can find a similar phenomenon emerging in the new works as well over the decade. The average age increased from 33.0 to 41.3 during 1978–92, and the average length of service from 13.4 to 21.6 years during the same period (Tekkō Rōren, *Rōdō Handbook* 1992). The average age and length of service has increased considerably and we can see that the advantage of the new works in this area had reduced greatly by 1992.

This trend of increasing average age and length of service in both the old and new works was a major incentive to introduce the new *Nōryokushugi Kanri* management system. This can be thought of as a sign of Japanese convergence towards British employment practices, in terms of a move from collective and age-based orientation to one placing more emphasis on individual and ability-based elements. Yet job demarcation in the British fashion does not exist in Japan and jobs are assigned more flexibly, so the convergence is only partial. The introduction of assessment of individual ability is a slight modification of traditional practice, but could be regarded as a sign of change in the direction of individually-oriented practice.

Table 6.10 Average age and average length of service of employees in old and new large-scale steel works (1978)

	Old large-scale steel works (Yawata)	*New large-scale steel works (Kimitsu)*
Average age	40.6	33.0
Average length of service (years)	20.5	13.4

Source: Tekkō Rōren, *Rōdō Handbook* 1978

MANAGEMENT AND LABOUR IN THE MAIN PRODUCTION PROCESSES

The main production processes are more affected by automation than anywhere else in the steelworks. Here, the level of automation and its impact on management and labour will be examined in relation to the ironmaking, steelmaking and rolling processes. Automation has standardised the content of operative jobs, changed the nature of skills and shortened the required period of training. A new type of management is required to administer workers in an automated process. A shift in management from that of collective and experienced-based skills to individual and knowledge-based skills can be observed in such workplaces.

Ironmaking process

The blast furnace plant in the ironmaking process has the highest number of installed computers; automation here is extensive and sophisticated (see Chapter 5, pp. 79–84). The specific automated functions are operation and control of the blast furnace; control of the hot stove; weighing of raw materials; reporting status of functions; and controlling the bottom of the blast furnace and slag removal. The above functions are integrated into the blast furnace process control system.[12]

The main worker teams in the blast furnace plant consist of a blast furnace operation team and a blast furnace hob work team. Jobs in the blast furnace operation team are carried out by ordinary workers, senior grade workers and supervisors (group leaders and a foreman). The work involves jobs typical of automation, such as observation of the blast furnace process control instrument board, recording, telephone communication with the blast furnace hob work team, and periodic inspection of the blast furnace. The character of such labour is mainly observation with a low-grade experiential skill, but involves nerve fatigue and responsibility. The work itself requires little or no technical knowledge of computers, but just an operational knowledge of the terminal unit; a week is said to be sufficient to learn its operation (Kobe Steel, Interview).

Table 6.11 shows the workforce composition of two blast furnace operation teams (No. 1 and No. 2 blast furnaces) in a new large works. Workers are on a three-shift/four-team system and each work group has five workers headed by a group leader. There is one foreman for two such groups. The average age and length of service of the workers are 26.1 years old and 7.7 years respectively. They are all graduates of senior high school (18 years old at the time of graduation). The average age of ordinary workers is 22.3, that of senior grade workers 26.5, group leaders 31 and the foreman 39. The average length of service of ordinary workers is four years, that of senior grade workers eight years, group leaders thirteen years and the foreman nineteen and a half years. When we examine the promotion situation in relation to age and length of service we find that, save for one instance in the No. 2 blast furnace team, promotion corresponds with the age and length of service of the workers.

Table 6.11 Workforce composition of blast furnace operation team in modern steel works (1977)

	Job rank	*Name*	*Age*	*School career*	*Length of service (years)*
No. 1, 2 B. F.	Foreman	A	39	High School	19.5
	Group leader	B	32	"	14.5
	Senior grade job	C	27	"	8.5
No. 1	Ordinary Job	D	24	"	6.5
blast		E	21	"	2.5
furnace		F	18	"	0.5
	Group Leader	G	30	"	11.5
	Senior grade job	H	26	"	7.5
No.2	Ordinary job	I	28	"	9.5
blast		J	22	"	3.5
furnace		K	21	"	1.5

Source: Kakogawa works (Kobe Steel): Interview

Table 6.12 Workforce composition of blast furnace hob work team in modern steel works (1977)

Job rank	*Name*	*Age*	*School career*	*Length of service (years)*
Foreman	A	51	Primary School	24.5
Group leader	B	34	High School	15.5
Senior grade job	C	35	"	13.5
	D	35	"	12.5
Middle grade job	E	32	"	8.5
	F	26	"	7.5
	G	29	"	7.5
Ordinary job	H	24	"	5.5
	I	21	"	2.5
	J	19	"	0.5
	K	19	"	0.5

Source: Kakogawa works (Kobe Steel): Interview

The jobs of the blast furnace hob work teams are carried out by ordinary workers, middle grade workers, senior grade workers, and supervisors (a group leader and a foreman) (Table 6.12). These jobs are tapping of pig iron, handling of hot iron and slag, sampling, and arrangement and repair of casting channels. Sampling is a job for beginners and requires from one to two weeks' experience; handling of hot iron requires three months' experience; pig iron tapping requires from two to three years' experience; and for dealing with emergencies, from three to five years' experience is required. Concerning job assignments, the foreman is not in charge of any actual jobs but oversees operations and provisioning of auxiliary materials. The group leader supervises workers; senior grade workers oversee actual work; middle grade workers are in charge of pig iron tapping and maintaining casting channels; and ordinary workers are in charge of hot iron and slag handling, sampling and repairing casting channels and arranging for tapping pig iron. These jobs require experience and skills of the old non-automated type and are noted for being hard work in high temperatures. No technical knowledge of computers is necessary. Job rotation exists within the teams but not with workers of the blast furnace operation team (Tekkō Rōren 1977).

The workforce composition shown in Table 6.12 therefore reflects labour characteristics different from those of the blast furnace operation team. For example, the average age is 29.5, 3.4 years higher than the blast furnace operation team. The length of service is 8.9 years, 1.2 years longer than the blast furnace operation team. The age of the foreman is remarkably high, 51 years old, which is twelve years older than that of the blast furnace operation team foreman. The age of the group leader is 34, two and four years older than the operation team group leaders. The age of senior grade workers is 35, eight or nine years older than those of the operation team. The age of ordinary workers ranges from 19 to 24, which is not much different from the operation team, but there are middle grade workers who are older and more experienced than ordinary workers, and as a result the number of job ranks in this team is larger than in the blast furnace operation team. In this team the age and the length of service correspond with promotion, and the relation between the two is more closely based on experience and skill than in the blast furnace operation team.

Steelmaking process

The LD plant in the steelmaking process is as advanced in automation as the blast furnace and rolling plants (see Table 5.4). Automated functions are control of the ending of blowing; control of steel quality and ingot composition; controlling standard operating procedures and molten steel properties; operating control settings of machinery (e.g. the height of lance, the amount of alloy); production control and information process; and compiling technical tables. All of these are integrated in the LD process control system.

Jobs in the LD plant are carried out by ordinary workers (with no distinction among them), senior grade workers and supervisors (group leaders and foreman). The jobs of ordinary workers are operation of gas recovery equipment, tilting the converter, charging auxiliary materials, motioning for cranes, preparation of alloy, etc. and sampling; jobs of the senior grade workers are measuring temperature while blowing, overseeing charging auxiliary materials, and adjustment of waiting time; and group leader jobs are blowing and overseeing the ending of blowing. The decision-making role of the group leader is said to require at least ten years' experience, relatively long compared to other jobs in the automated process. These other tasks are all typical of automation and are carried out in the operation room. They consist of observation of instrument boards, recording, inspection, and communication. Among ordinary workers, there is job rotation. Computer-related knowledge necessary for the immediate work only involves operating the computer terminals (Tekkō Rōren 1977).

Table 6.13 shows the workforce composition of an LD plant operation team. The workers are organised in the three-shift/four-team system and there are six workers headed by a group leader in the team. The age of the ordinary workers is from 24 to 25, slightly older than the members of the blast furnace operation team (22.3 years old) and the blast furnace hob work team (20.8 years old). The average age of senior grade workers is 32, slightly older than the blast furnace operation team and slightly younger than the blast furnace hob work team. The age of group leaders ranges from 35 to 36, slightly older than both the blast

Table 6.13 Workforce composition of LD converter operation team in modern steel works

Foreman —— Group Leader —— Senior Grade Job —┬ A (Ordinary Job)
├ B (Ordinary Job)
├ C (Ordinary Job)
└ D (Ordinary Job)

	Foreman	*Group leader*	*Senior grade job*	*Ordinary job*
Age	*45-46*	*35-36*	*32*	*24-25*
Content of job	Management of work	Supervision of work, blowing, blowing out	Taking temperature while blowing, direction for putting in supplementary materials, adjustment for waiting time	Operation of gas collection plant, tilting converter, putting in supplementary materials, arrangement of Ferro alloy, sampling

Source: Tekkō Rōren 1977: 50

furnace operation team and the blast furnace hob work team. The foreman, at 45–46, is older than the blast furnace operation team and younger than the blast furnace hob work team. Almost all workers are graduates of high school. In terms of age and promotion, the two factors correspond with each other without exception.

Rolling process

Automation is also high in the hot rolling plants in the rolling process, which has the next largest number of computers after blast furnaces (see Table 5.4). Automated functions are setting the heating furnace; mill paging; setting the rough rolling equipment; adaptation control; setting automatic gauge control; setting coiler and measuring instruments, and compiling technical tables. These are all integrated into the hot rolling process control system.

The main jobs of the hot rolling operation team are carried out by ordinary workers and supervisors (group leader and foreman). Ordinary workers carry out rough rolling and finish rolling (splay control, speed control, flying shear, levelling, and inspection). The job of speed control, which is directly related to the width and thickness of slab, and the control of forms of rolling stands are the most difficult jobs; and rough rolling is the second most difficult job, requiring from two to three years' experience. Splay control and inspection are work for beginners. The character of work in this team is also typical of automated work, consisting of observing instrument boards, recording, communication and inspection. Job rotation exists among ordinary workers and promotion is limited within this work team, where the foreman is the highest rank. The computer-related knowledge required for the immediate work is limited to the operation of computer terminals (Tekkō Rōren 1977).

Table 6.14 shows the workforce composition of a hot rolling mill operation team. Workers are on the three-shift/four-team system, a single team being composed of six ordinary workers and a group leader. The age of ordinary workers ranges from 20 to 30, and in terms of maximum age is slightly older than the blast furnace operation, blast furnace hob work and LD converter operation teams. The age of group leaders is between 30 and 35, almost the same as that of the other teams. The age of the foreman is over 35 years, slightly younger than the above-mentioned teams. The workers in this team are also almost all high school graduates.

Promotion in this case too corresponds with age and length of service, but as there are only three positions, including foreman, in this team,

Table 6.14 Workforce composition of operation team for hot rolling

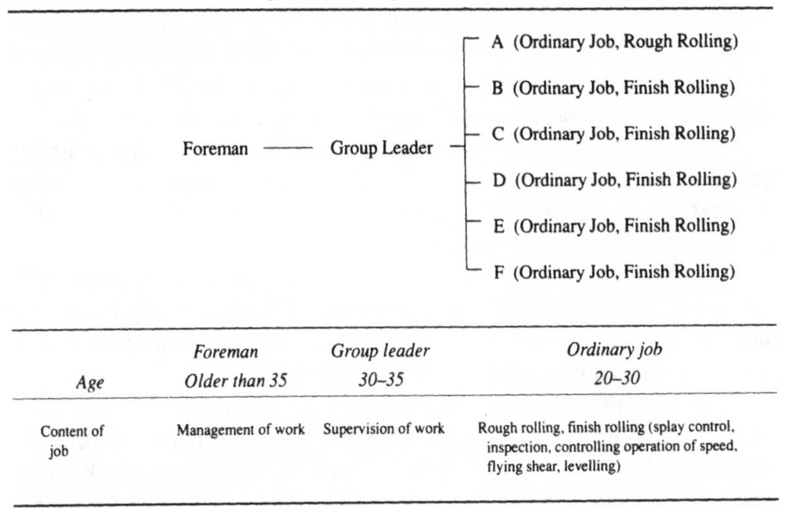

	Foreman	*Group leader*	*Ordinary job*
Age	*Older than 35*	*30–35*	*20–30*
Content of job	Management of work	Supervision of work	Rough rolling, finish rolling (splay control, inspection, controlling operation of speed, flying shear, levelling)

Source: Tekkō Rōren 1977: 51

we see a larger age range among ordinary workers than in the other groups.

Management at the workplace

We can summarise the influence of automation in the main processes upon management and labour as follows.

First, there exist differences in the character of jobs due to differences in the technology used in production, between knowledge-oriented 'automated jobs' and experience-oriented 'skilled jobs' in the non-automated process. The former are found in the blast furnace operation, LD converter operation and hot rolling mill operation teams, and the latter in the blast furnace hob work team. The job character in the former operation teams is typical 'automated work', consisting mainly of observation of instrument panels, which can lead to stress-related nerve fatigue, while the job of the blast furnace hob work team is a heavy physical job in high temperatures using various tools and machines. Thus the difference in type of job corresponds with the level of automation. The difference in the job character is further related to management, as reflected in the number of ranked positions. Namely, the number of ranked positions in the blast furnace hob work team, where the work is based on one's level of experience and skill, is five, while the operation teams have fewer positions, either three or four. Among these, the average age and length of service of each position rank shows similar trends although there are slight differences between the teams of the different plants. In particular, the similarity in job character and average age and length of service suggests that the jobs and skills in the different work teams are becoming increasingly similar, resulting in the content of labour itself becoming much the same through the equipment modernisation resultant from the high level of automation.

Second, a sophisticated knowledge of computers is not required in the actual work. This suggests that the technological theory behind the huge automated equipment is complex but the operation of the equipment itself is standardised as well as simplified.

Third, length of service, which shows the degree of experience and skill, is not long. The average length of service of ordinary workers who undertake the actual work is about 7.6 years. Provided that a high school graduate continues to work until retirement (until the age of 55), it will become thirty-seven years. When we consider this, we see that the present average length of service of ordinary workers is short, suggesting that the period of skill formation is reduced as an effect of equipment modernisation. In addition, the work of the blast furnace hob work team, so far considered to require experience and skill, has also become standardised and simplified to the extent that it can be done after half a year's experience, except for the tapping of pig iron, which is considered even today to require from two to three years' experience. The work of supervisors (group leader and foreman), whose length of service is relatively long, is to judge unusual indications appearing on the instrument board and decide how to deal with them. What they have charge over, therefore, are things like judgement, decisions, instructions in case of emergency, and communicating with maintenance and other necessary sections.

Fourth, there exists a conflict between standardised and simplified 'automated job' and education. All workers, except some foremen with long length of service, are high school graduates, indicating that educational levels have become high. But this does not mean that the 'automated job' in the operation teams and the job in the blast furnace hob work team require high school level knowledge and skill; it rather means that it has become difficult to recruit junior high school graduates (15 years old) because advancement to high school on completion of junior high school has come to account for more than 90 per cent of students. Another reason is that companies are not allowed legally to use workers at the age of 15 on the three-shift/four-team system. The tasks of ordinary workers do not include jobs requiring judgement and have become standardised and simplified to the extent that high school graduates are not necessary. This is shown by the nature of the 'automated job' and the average age and length of service; in addition, the qualification for employment is not limited to technical high school graduates, but includes graduates from ordinary high schools, commercial high schools, and agricultural high schools (Tekkō Rōren 1977).

Fifth, job rotation exists within the same work team. This is done in order to alleviate mental fatigue, which can become a cause of stress and dissatisfaction of workers in an automated production environment with large-sized equipment and high-speed operation. The main job of workers with a high school education has become simplified and standardised observation work, and rotation helps to raise the morale of workers and aims to reduce the number of workers and bring about a more intensive use of labour in the 'highly efficient small number team'.

Sixth, seniority ranking at the workplace is maintained under the life-time employment practice. The possibility of all workers in a team being promoted is extremely slim; however, among those who are promoted, the promotion is based upon age and length of service, thus showing that seniority ranking on the basis of life-time employment, even with plant and equipment modernisation, is still maintained. Moreover, the promotion route – from ordinary to middle grade position, senior grade position, group leader and eventually foreman – is narrowly limited within the team. Therefore we may assume that workers are doing their daily job in a small, competitive, and yet co-operative, work team.

Shokunō Shikaku Seido[13]

In the hot rolling operation team, where workers do their job at a high level of automation, the average age of workers is younger than in others. In the team, seniority ranking is maintained, but in terms of the steel works as a whole it is modified by *Shokunō Shikaku Seido* (Job Qualification System). Thus, younger workers may reach higher positions; the ages of group leaders and the foreman of the hot rolling operation team are lower than in other teams.

In terms of job character, skill and employee training, the most common job in the main production processes has become the operation of automated equipment. In contrast to the high level of education workers had in school, the job of ordinary workers is standardised and simplified. Therefore, the period of skill formation has become much shorter than in the past. For example, Table 6.15 shows the results of a survey examining the length of time necessary to learn a job in the new large-scale works. The percentage of jobs which require more than five years is only 6 or 7 per cent, while the percentage of jobs which are able to be mastered within a year accounts for more than 70 per cent. As already noted, the average length of service of ordinary operation workers is about 7.6 years, so it means that almost all workers have been in their present job more than the period of time required to master the job. This suggests that their high potential ability acquired by higher education is kept at a low level in their daily work, a situtation which could lead to problems of worker satisfaction. Job rotation within the same workplace has become possible due to standardisation, simplification of labour, and shortening of the skill formation period. This has the good effect of developing the morale of workers, lessening labour alienation and reinforcing labour management. A new attempt to develop job rotation beyond the same workplace is planned, especially the assignment of operatives to daily and regular equipment maintenance work aimed at labour rationalisation and improved morale (Tekkō Renmei, *Tekkō no IE* 1979: No. 6).

Shokunō Shikaku Seido is used in labour situations affected by automation as an effective management tool to accommodate difficulties arising from the principle of seniority-based promotion and the effects of uneven development of automation. Though automation is at a high stage in the main production processes, it is not consistent within them. This is reflected in the number of promotion positions. In a new plant, workers are promoted earlier and at a younger age than in an old plant. This makes it difficult to apply a uniform seniority system based

Table 6.15 Years required to learn operators' jobs in modern steel works (Nippon Steel, Nagoya works)

Factory A (surveyed 56 jobs)		*Factory B (surveyed 17 jobs)*	
Years required to learn jobs	*Percentage of jobs to be learnt (%)*	*Years required to learn jobs*	*Percentage of jobs to be learnt (%)*
1	79	1	70
2	5	2	12
3	9	3	12
5	7	5	6
Total	100.0	Total	100.0

Source: Tekkō Renmei, *Tekkō no IE* 1979: No. 6

on age and length of service throughout the works and company. In the hot rolling plant, the age and length of service of those who are promoted to supervisory positions are lower than in the blast furnace and LD converter plants. This is because the expansion and increase of the rolling mills have enhanced promotion opportunities, and furthermore, hot rolling plant workers are given higher qualifications based on an evaluation and exam system in *Shokunō Shikaku Seido*. In some older plants, even though the age and the length of service are greater, workers are positioned at lower qualification levels, and as a result promotion to supervisory positions does not occur.

EDUCATION AND TRAINING FOR PRODUCTION WORKERS AND FOREMEN

In-firm education and training becomes necessary in general when the demand for labour exceeds that of supply and also when the economy is expanding through technological innovation. The steel industry in the period under study was experiencing such a corporate environment and this incurred the need to create a firm-specific labour force through developing its own education and training.

The percentage of production workers in terms of total employees has always been high, as already examined, and the level of their technical knowledge and attitudes plays an important role both in terms of productivity and quality in production. As the production process in modern steelworks requires continuous, high-speed, automatic operation of equipment twenty-four hours a day, it is vitally important that workers have high morale and faithfully carry out their jobs under the instruction and control of the works management. For this reason, the education and training of production workers in both technical knowledge and experience, as well as in their attitudes and minds (particularly the ideological aspect of inducing company loyalty) is crucial – it is an essential precondition for 'harmonious' industrial relations and becomes an important means of labour management.

Education and training and its relation to the promotion of production workers were formulated in the early 1960s and have only been slightly modified since. This will be examined in the case of Yawata works, which is highly representative for its systematic programmes. New recruits at this time were already almost all high school graduates due to the spread of high school education, except that specific occupations, such as shunting, were assigned to junior high school graduates (16 years old at the time of graduation). The education system for workers shown in Figure 6.1 comprises, first, an introductory education, and a foreman education at its highest stage. In between there are on-the-job training programmes and technical knowledge and theory programmes by turns. The average number of years required for promotion into each subsequent rank are three, ten and twenty years respectively. These promotion periods correspond with the education and training

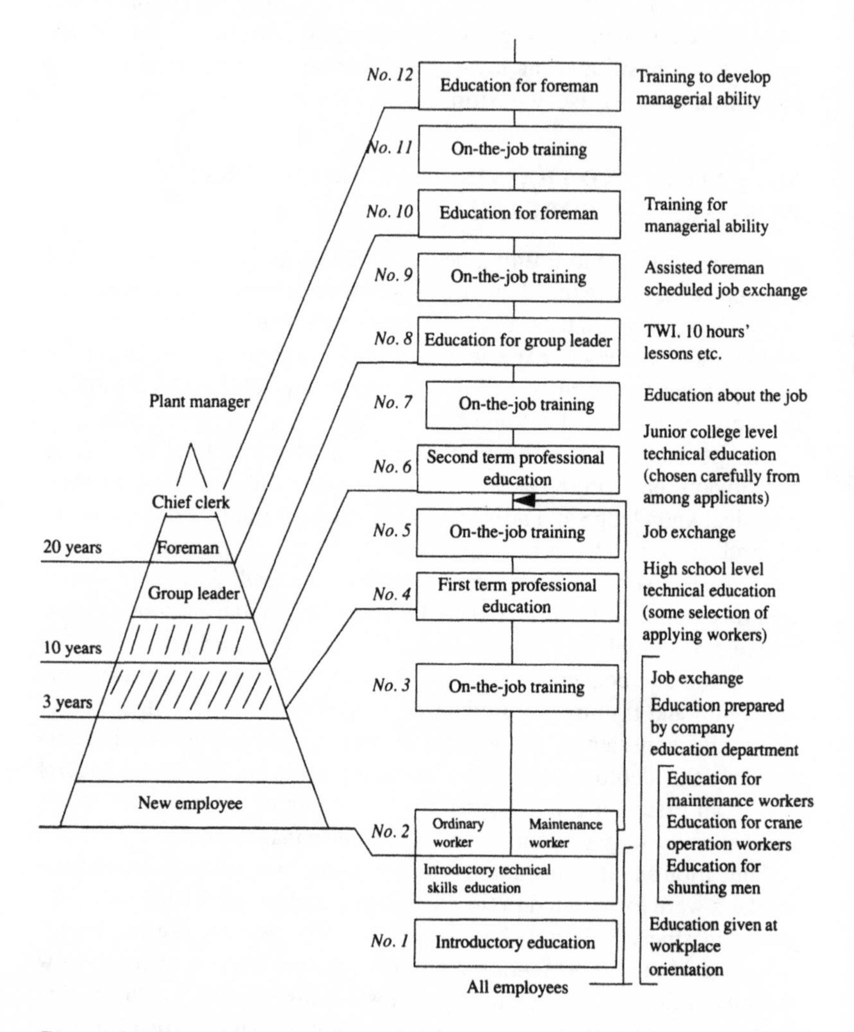

Figure 6.1 Education system for technical personnel (workers) (Yawata Steel Corporation)

Source: Tekkō Renmei 1962: 9

Note: The above figure is one example: there exists some difference between departments programmes. Promotion is thus reinforced and legitimised by the level of education and training.

First, introductory education and introductory technical education is provided for every new employee; the latter is divided into two sections, one for production workers and the other for service workers, crane drivers and shunting workers. The purpose of this education is 'to give knowledge and safety training necessary for undertaking the job assigned to the new employee' (Tekkō Renmei 1962: 15) and the result of the assessment of each worker during the education period is reflected in their promotion.

Second, the first term professional education is given to those who are assessed highly, have passed a test, and in addition have more than five years' length of service in the same workshop. The purpose of this education is to give 'a professional theoretical knowledge of the work which is difficult to obtain through on-the-job training for workers who have already acquired experience and skill of more than a certain period' (Tekkō Renmei 1962: 17).

Third, the second term professional education is given to certain workers selected by assessment and examination, with over ten years' service. The aim of this education is 'to give a higher professional theoretical knowledge (junior college level) to workers of medium standing who have finished the first term professional education' (Tekkō Renmei 1962: 17). The assessment of each worker during the education period is reflected in their promotion.

Fourth, the aim of group leader education is 'to give elementary managerial methods to those who are newly appointed as a group leader' (Tekkō Renmei 1962: 17).

Fifth, the aim of foreman training education is 'to give knowledge of the operation of the plant and managerial methods necessary for first-line managerial personnel to the foreman candidates' (Tekkō Renmei 1962: 19). The applicants are chosen from those receiving high evaluation in the plant, a recommendation from the plant manager and the consent of higher labour personnel and a personnel manager. The result of this education will be assessed by an examination and report; those who have finished this education may become foremen.

Sixth, the study and training for foreman aims to maintain and improve the quality of foremen. Every new foreman must attend after a certain period in the post, and the education includes various managerial techniques such as IE (industrial engineering) and QC (Quality Control), basic understanding and methods of HR (Human Relations), PR (Public Relations), industrial psychology and TWI (Training Within Industry).

The length of education and training in this period is shown in Table 6.16. The technical skill education needed for the daily work is comparatively short, even taking the year-and-a-half maintenance education into account. In order to receive the professional education, one must have a certain length of service, a good evaluation from supervisors, and pass an examination on certain subjects. This education is an essential for promotion to a supervisory position. The qualifications required to be chosen as a candidate for foreman are very strict. One must be qualified in every aspect as a first line managerial staff person and must be approved by superiors of the labour and personnel departments. Foreman education includes industrial psychology as well as the various managerial techniques.

Table 6.16 Education and training period for steel workers, early 1960s (Yawata Steel Corporation)

Education and training	*Period of education and training*
Introductory education	10 days (full-time)
Elementary technical education	Ordinary worker 1 week to 1 month (full-time) Equipment worker 1.5 years (full-time)
Primary professional education	3–5 months (part-time)
Secondary professional education	about 1 year (part-time)
Education for group leader	about 15 days (2 hours x 5 days x 3)
Education for future foreman	about 2 months (full-time)
Education for foreman	about 6 days (full-time)

Source: Tekkō Renmei 1962: 15–9

The educational breadth of the foreman reflects the importance of his functions as a local representative of management in both production control and labour control. Thus, both the education and training system and promotion are geared for progress along the same route to foreman level.

In the 1960s the line and staff system, the foreman system, standardised work methods set by the IE department, the recruitment of high school graduates and the introduction of *Shokumukyū* (pay according to job function) were introduced, but the 1970s witnessed the growth of *Nōryokushugi Kanri*, composed of *Shokunō Shikaku Seido* (Job Qualification System) and *Jishukanri Katsudō* (Autonomous Management Activity), and employee education and training was also reorganised so as to be incorporated in *Nōryokushugi Kanri* (see Figure 7.2, p. 130).

Nōryokushugi Kanri differs from the 1960s' management system in two respects. First, education and training is directly tied to the position held; and second, it consists of three different programmes. Among these, the hierarchy of education, which is immediately related to the *Shokunō Shikaku Seido*, the core of the *Nōryokushugi Kanri* developed in the 1970s, is considered to be the most important tool of labour management. In the hierarchy of education, promotion to a higher qualification is by means of personal assessment and interview in case of ordinary production workers and also an examination in the case of workers with qualifications above *Shuji*. Those promoted attend a programme prepared specifically for each qualification; these programmes include overnight study and training retreats (Table 6.17). The programme for foreman candidates (two months) is similar to the foreman nurturing programme (two months) of the early 1960s; the education and training for foreman remained most important in the 1970s, and is a major focus of workplace management. Ordinary workers are also given a programme which includes overnight study and training retreat, assumed to incorporate them effectively into the whole company education and training programme.

The special professional education is composed of four programmes, starting from the basic subjects study and ending with a special professional subjects study programme (*Senka*). It aims at teaching the

Table 6.17 Period of education and training by qualification (Nippon Steel, Kimitsu works)

Education by qualification	*Period of education and training*	
Education for new employees	Introductory education	4 nights 10 days
	Joint education of new employees and coaches	1 night 2 days
Education for the person in charge	Education for young workers	1 night 2 days
Education for instructors	Education for coaches	1 night 2 days
	Education for instructors	3 nights 5 days
Education for chief persons in charge	Education for leaders	1 night 3 days
Education for future group leader	Education for assistant group leader	2 nights 7 days
Education for group leader	Primary education	2 nights 4 days
	Secondary education	3 nights 4 days
Education for future foreman	Education for future foreman	2 months
Education for foreman	Primary education	2 nights 3 days

Source: Tekkō Renmei 1977: 12–14

theory of on-the-job training as well as special technical knowledge and theory. The special professional subjects study programme continues for about four months, and participants concentrate on studying, being completely separate from the job. This is the only path to becoming a foreman. These programmes are equivalent to the first two years of liberal arts and technical subjects at university level. The age of those eligible ranges from 35 to 40. The basic subjects study programme and the special professional subjects study programme Nos. 1 and 2 of Kimitsu works of Nippon Steel Corporation shown in Table 6.17 correspond to the introductory technical education, the first term professional education, and the second term professional education of the early 1960s shown in Table 6.16. In comparing the length of education for ordinary workers of the early 1960s with that of the mid-1970s, the former was one year and a half, while the latter was about half a year, showing a great reduction. On the other hand, the education and training for foremen was greatly extended; in the early 1960s it was about two months of day-time education, while in the mid-1970s it had become a two-month programme for foreman candidates in addition to a four-month day-time special professional subjects study programme (*Senka*). As to the maintenance-related study and training programme of the mid-1970s (Table 6.19), the time period for both basic maintenance and basic engineering programmes was nine months, down considerably from the one-and-a-half-year introductory technical education for maintenance workers of the early 1960s (Table 6.16). The longer term educational programme was replaced in part by separate short technical education and training programmes.

Other study and training programmes considered to be important in *Nōryokushugi Kanri* are the programme for the leaders of *Jishukanri Katsudō* (Autonomous Management Activity) and the safety programme. The former is for leaders in small management groups, such as QC and ZD (Zero Defect Movement) and various self-motivating activities.[14] The leaders are usually foremen or leaders of work teams; they try to find ways of reducing costs and also of raising work morale and loyalty to the company. The safety programme has become important in every works because in an environment of large, high-speed, continuous, and automated equipment, even a minor accident in the works may not only be dangerous for an employee, but will also disrupt the continuous operation, thus harming the profitability of the enterprise.

Thus, in the 1970s, the time spent on special technical education for ordinary workers was much shorter than in the 1960s, while the educational period for foreman candidates was extended. The foreman's technical and managerial knowledge and techniques were enhanced and enriched, establishing a firm foundation for reinforced management of workers by the foreman. The reduction of the period for special technical education and training given to production workers suggests a lowering and levelling of the character of labour and technical skills due to plant and equipment modernisation and the development of automation. This has developed the division and diversification of labour between ordinary workers and technicians and engineers, and at the same time created technical and economic conditions promoting the sub-contracting of highly technical maintenance and service work (Ishida 1967). Education and training was a means towards the *Shokunōshikaku Seidō*, where its hierarchy was considered to be important; this implies that education and training play an important role in reinforcing hierarchical labour management. Moreover, as is very apparent in *Jishukanri Katsudō*, importance is also attached to small group activity (Sohyō, *Chōsa Geppō* 1974: No. 96).

CONCLUSION

The new production system of comprehensive high automation within the steelworks has made it possible for management to restructure the workforce and introduce measures designed to foster efficiency and loyalty among their staff.

There are some differences in the extent of division of labour between old and new steelworks, especially in sub-contracting, reflecting environmental influences, but over time we find convergence occurring between them.

In-firm education and training was an integral part of organic compatibility in the 1960s, together with traditional management concepts based upon a principle of collectivity. However, in the 1970s the *Nōryokushugi Kanri* system was introduced to iron out certain dysfunctions in existing systems and was adopted overall. It represents a conceptual shift in Japanese management from a collective to an individual approach to its staff. The in-firm education and training has thus been an important instrument of management at both the positive and contrived compatibility stages in the Japanese steel industry.

7 Management and labour at head office (Japan)

During the growth period of 1960–73, management functions throughout the industry became concentrated at head offices and a corresponding complexity in the division of labour there was also developed. The head offices themselves expanded as single-unit organisations without any diversification in corporate structure. Another important development was the institution of thorough education and training for employees at head office, and this system came to be used as a device for personnel management. This chapter examines the change in management and labour at head office which ensued from the modernisation programmes.

HEAD OFFICE AND TOP MANAGEMENT

Our examination of head office and top management begins with the analysis of white-collar employees and their concentration at head office. If we compare the ratio of white-collar employees to blue-collar employees in 1960 (Yawata Steel Corporation) and 1978 (Nippon Steel Corporation), we notice a change from 19.5:80.5 per cent to 30.1:69.9 per cent. Thus the ratio of white- to blue-collar employees increased more than 10 per cent (Yawata Steel 1961; Nippon Steel, *Yukashōken Hōkokusho* 1978).

As white-collar staff are composed of clerical, technical and managerial personnel at head office, works and other plants, the increase of white-collar employees needs to be examined in relation to other changes in the corporate organisation. The white-collar percentage at head office as a proportion of total personnel is shown in Table 7.1. Head office staff increased by about 2.5 times from 1960 to 1978. Such an increase suggests a rapid concentration and large increase of managerial functions at head office level. This takes the form of extensive division of labour as well as socialisation (interdependence) among clerical, technical and managerial personnel.

The increase in head office staff is a common characteristic of the companies which expanded at this time, and in the case of steel this was due to the rapid and continuous increase in steel demand. Due to the development of on-line and comprehensive data processing systems needed for large-scale, continuous, high-speed and automated production, the previous sharing of managerial functions by both works and head office was thoroughly rationalised as well as specialised and this has as a result defined steelworks as solely production units under strong central control. It is due to this organisational requisite that the size and percentage of head office staff increased (Nippon Steel 1978b).[1]

Table 7.1 Workforce composition of large steel companies by head office and other establishments

	1960 (Yawata Steel Corp.)		*1978 (Nippon Steel Corp.)*	
	No. of employees	*Ratio (%)*	*No. of employees*	*Ratio (%)*
Head office	1,371	3.4	6,564	8.6
Other establishments	39,089	96.6	69,470	91.4
Total	40,460	100.0	76,034	100.0

Sources: Yawata Steel 1961; Nippon Steel, *Yukashōken Hōkokusho* 1978

The head office employee composition is compared at two points, 1960 and 1978, in Table 7.2. Note that the number of directors, managerial staff (managers and section chiefs), and ordinary staff have all risen. The rise in managerial staff is especially high both in number and percentage. This reflects a concentration of the managerial function of the company to head office level, resulting in the greater control of head office over other establishments – in other words, the centralisation of power at head office.

Top management increased remarkably at head office, as shown in Table 7.3, and we can see that the number of top management personnel officially titled *Torishimariyaku* (director) increased from twenty-two in 1961 to fifty-two in 1978.[2] Positions which did not exist in 1961 but did in 1978 were chairman, i.e. chairman of the board as well as representative director, and standing auditor, whose responsibility is, in theory, to keep the board accountable (in practice, however, they are usually chosen from those who have not been promoted to core members of the board). There was a five-fold increase in vice presidents, being executive vice presidents as well as directors, and managing directors more than doubled. The above increase in each position of top management has created a shift of relative importance of those who are above managing directors, whose percentage rose to 49.9 per cent in 1978 from 40.9 per cent in 1961. The management policy council, which consists of chairman, president, and vice president, has become the most important decision-making body as well as consensus creation forum of the top management.[3]

Table 7.2 Head office employee composition of large steel companies

	1960 (Yawata Steel Corp.)		*1978 (Nippon Steel Corp.)*	
	Number	*Ratio (%)*	*Number*	*Ratio (%)*
Directors	22	1.6	52	0.8
Managers and section chiefs	97	7.1	813	12.4
Lower rank and ordinary staff	1,252	91.3	5,699	86.8
Total	1,371	100.0	6,564	100.0

Sources: Yawata Steel 1961; Tekkō Shimbun-sha 1960; Daiyamondo Publishers 1979; Nippon Steel, *Yukashōken Hōkokusho* 1978

Notes: (a) Managers are inclusive of all managers higher than section chiefs and excepting directors

(b) The number of directors and managers of Yawata Steel Corporation, as of June 1960, and the number of lower rank and ordinary staff of the same corporation as of November 1960. The number of directors and managers of Nippon Steel Corporation, as of August 1978, and the number of lower rank and ordinary staff of the same corporation, as of March, 1978

Table 7.3 Growth of top management in large steel companies

	1961	*1978*
Chairman	0	1
President	1	1
Vice President	2	10
Managing Director	6	14
Director	11	22
Standing Auditors	0	2
Auditors	2	2
Total	22	52

Source: Yawata Steel, *Yukashōken Hokōkusho* 1961; Nippon Steel, *Yukashōken Hōkokusho* 1978

One of the most distinctive features of the top management structure and function is that all members are concurrently members of the board, which is in theory an organisation for corporate governance, and of the executive, who are responsible for carrying out what the board has decided. This dual responsibility is reflected in titles such as 'director and executive vice president'. Hence, no distinction is made in function between corporate governance and execution, either in terms of organisation or people. Decision-making will usually be done collectively in the above management policy council. Here again, decision-making and execution is shared among top management staff.[4]

In Table 7.4 we examine relations between the positions and responsibilities of top management. Among the ten vice presidents in 1978, five were in charge of general management, and the other five in charge of divisional management. Thus, those top management vice presidents who are also directors bear specific responsibilities of execution. The general management areas assigned to the first five positions are business planning, management of overseas activities, management of sub-contract affairs, general research, management of environment, affiliated activities, labour, general affairs, information systems, accounting, finance, sales, engineering business, raw material and fuel, machinery, and technology. It is also important to note that each position is responsible for a number of different managerial functions. Areas of divisional management for the remaining five are planning and development headquarters, new large-scale works (Kimitsu, Nagoya, Oita), and old large-scale works (Yawata). In both general and divisional management it is clear that the predominant importance of top management is in the function of production and its development, rather than finance or anything else, which characterises the stage of organic compatibility.

Of the fourteen managing directors, eight are in charge of general management and the other six are in charge of divisional management. The areas assigned to general management are business planning, management of overseas activities, management of sub-contract affairs, affiliated activities, labour, general affairs, information systems, accounting, finance, sales, production control, technology for steel piping, energy control, equipment, secretarial office, personnel, faculty development (education and training), sales management, export, and technology services. The areas assigned to divisional management are engineering business, R & D headquarters, New York office, new medium-sized works (Sakai) and old medium- or small-sized works (Hirohata, Muroran).

Of the twenty-two ordinary directors, four are in charge of general management and eighteen are in charge of divisional management. The areas of general management are general research, management of environment, raw material and fuel, and production control. The responsibility of these ordinary directors is to assist director and executive vice presidents and/or managing directors. The areas included in the divisional management are headquarters of engineering business, headquarters of R & D, technology and equipment centre, headquarters for co-operation with China, Osaka office, new large-scale works (Kimitsu), old large-scale works (Yawata), old medium or small works (Muroran, Kamaishi, Hikari), factory (Tokyo), personnel, production control, labour, finance, and management of sheet sales.

Table 7.4 Division of function in top management of large steel companies

	1961		*1978*	
	General Management	*Divisional Management*	*General Management*	*Divisional Management*
Chairman				
President				
Vice president	1	1	5	5
Managing director	4	2	8	6
Director	2	9	4	18
Standing auditors				
Auditors				

Source: Yawata Steel, *Yukashōken Hōkokusho* 1961; Nippon Steel, *Yukashōken Hōkokusho* 1978

Directors are structured in three tiers and two areas of function in both periods; 1978 represents a quantitative expansion of the situation pertaining in 1961. The majority of ordinary directors are in divisional responsibilities, while the majority of managing directors are in general management. Vice presidents are spread equally over two functions in both periods. Ordinary directors in general management were acting as assistants to managing directors and managing directors to vice presidents. This all created a hierarchy of function as well as responsibility in the top management structure.

In 1978, top management – the area of general management, where three positions (vice president, managing director and ordinary director) overlap – oversaw the organisation of business planning. It is thereby suggested that this organisation is the most important and influential in the area of general management. There are ten organisations in which two positions (vice president and managing director) overlap: management of overseas activities, management of subcontract affairs, affiliated activities, labour, general affairs, information systems, accounting, finance, sales, and raw material and fuel. These organisations are therefore considered next in importance after business planning in the area of general management. Among these organisations, finance, labour and sales organisations are headed by ordinary directors as the divisional management representatives, and thus these three organisations are more important than the other seven listed. Moreover, organisations of which managing directors are in charge in the area of general management, but which overlap with the responsibilities of ordinary directors, are production control and personnel departments. These two departments are suggested to be third in importance.

As for divisional management, vice presidents are in charge of development planning headquarters, new large-scale works (Kimitsu, Nagoya, Oita), and old large-scale works (Yawata). It is suggested that this organisation and these works are the most important in the area of divisional management. Moreover, ordinary directors are also assigned to Kimitsu and Yawata works, suggesting that these two sites are the most important. Managing directors are in charge of engineering business headquarters, R & D headquarters, New York office, new medium-sized works (Sakai) and old small- and medium-sized works (Hirohata, Muroran), which are consequently given secondary importance in divisional management. Of these organisations, Muroran works, engineering business headquarters and R & D headquarters may be more important, since ordinary directors are also assigned to these organisations. Organisations of which ordinary directors are in charge without overlapping with higher level directors are the equipment technology centre, Osaka office, old small-sized works (Kamaishi, Hikari) and the Tokyo factory (tube manufacturing). As to the China co-operation headquarters, the vice president who is the head of development planning headquarters, the managing director who is the head of engineering business headquarters, and also one of the ordinary directors are assigned to it. It is apparent from this that Nippon Steel Corporation holds a positive attitude towards the People's Republic of China.

The major objective of corporate strategy in this period therefore affected top management function regarding production and its expansion.

According to analysis based upon the *Steel Year Book* 1960 and Daiyamondo *Company Personnel List* of 1979, the composition of managerial personnel (including section chiefs and higher positions) of the head office of these two cases (Yawata Steel and Nippon Steel) by time period and the relative organisational development is as follows.

The number of organisations increased from twenty to forty-nine between the two periods. Directors, including all top management figures, as a percentage of all managerial personnel declined between 1960 and 1978, suggesting that the managerial functions of top management conversely increased and head office functions were further centralised to top management. Organisations with a greater percentage of personnel in 1960 were the sales department (12.6 per cent), purchasing department (7.6 per cent), technology department (6.7 per cent), Tokyo Research Institute (5.9 per cent), Osaka office (5.9 per cent), general affairs department (5.0 per cent), accounting department (5.0 per cent), planning department (5.0 per cent), Nagoya office (4.2 per cent) and the fund department (3.4 per cent). General managers of the sales department, accounting department, planning department, and fund department served as ordinary directors of top management. Organisations with a greater percentage of personnel in 1978 were the equipment technology centre (24.4 per cent), R & D department (10.1 per cent), planning and development department (8.4 per cent), technology services department (3.9 per cent), general research department (3.4 per cent), production control department (3.4 per cent), technology department for steel pipe (2.4 per cent), sales department for construction material (2.3 per cent) and sales department for steel pipe (2.0 per cent). An ordinary director heads the equipment technology centre, a managing director heads the R & D department, a director and executive vice president the planning and development department and an ordinary director the production control department. In 1960, an ordinary director headed the personnel department, which accounted for only 1.7 per cent of total personnel. In 1978, ordinary directors headed not only the personnel department (1.2 per cent), but also the labour department (1.3 per cent), fund department (1.2 per cent), sales department of sheet steel (0.7 per cent) and Osaka office (0.3 per cent). This suggests that these organisations, though small in size, played an important role in managerial function. Thus managers of organisations with a higher percentage of total managerial personnel and also those with lower percentages, but having an important managerial function, are part of the top management class.

The characteristics of managerial personnel composition in 1978 can therefore be summarised in the following way:

1 The percentage of total personnel of a number of organisations declined compared with 1960 in the sales department and various branch offices. Meanwhile the percentage of total personnel increased in organisations concerned with technology and staff function, i.e. equipment technology department, R & D department, planning and development department and general research department.
2 The organisation concerned with labour management developed into five organisations: personnel department (1.2 per cent), faculty development office (0.5 per cent), management department of sub-contract affairs (0.5 per cent), production control department (3.4 per cent), and labour department (1.3 per cent), accounting for 6.9 per cent of total personnel. This indicates that labour management functions increased and diversified relative to the total managerial functions, due to a larger workforce which had grown in complexity, and the necessity to achieve its rational and efficient use.[5]

Managerial personnel composition and hierarchy at the head offices in 1960 and 1978 is shown in Table 7.5. The number of positions in 1960 was nine, rising to twenty in 1978. The percentage of total positions which were principal positions – director, manager, and section chief – were 18.5 per cent, 10.1 per cent and 46.2 per cent respectively in 1960, while they were 6.0 per cent, 5.0 per cent and 56.3 per cent respectively in 1978. Thus the percentage of directors and managers decreased, while that of section chiefs increased. A similar trend is seen if we compare the percentage of positions in the senior management class and the assistant to senior management class. The assistant to senior management post increased remarkably in number as well as in percentage of total positions, accounting for 10.1 per cent in 1960 and 22.5 per cent in 1978. By 1978 various new staff positions had been created in middle management: room chiefs, group leaders, assistant group leaders, head of laboratory, assistant head of laboratory, assistant chief of laboratory and assistant manager in charge, accounting for 8.7 per cent in 1978.[6]

This expansion and diversification of the head office organisation reflects an increase in managerial functions and an effect of seniority-based promotion. Managerial power became concentrated in the director positions, and higher decision-making functions for day-to-day operation became concentrated in senior managers. Middle management positions expanded, as did assistant to senior management class. Under the long-term employment and seniority-based promotion systems a gap between the number of necessary core (line) managers and the number of qualified candidates for the positions was created. Just

Table 7.5 Managerial hierarchy of head office in representative steel companies

	1960 (Yawata Corp.)			1978 (Nippon Steel Corp.)	
Managerial positions	*Number of personnel*	*Ratio (%)*	*Managerial positions*	*Number of personnel*	*Ratio (%)*
Top management			*Top management*		
Director	22	18.5	Director	52	6.0
Advisor	5	4.2	Advisor	3	0.3
Sub-total	27	22.7		55	6.3
Senior management			*Senior management*		
Manager	12	10.1	Assistant manager to head-dept.	2	0.2
Head of office	7	5.9			
Secretary	1	0.8	Manager	43	5.0
			Specialist manager	13	1.5
			Room chief	12	1.4
			Head of R&D Institute	2	0.2
			Assistant head of R&D institute	1	0.1
Sub-total	20	16.8		73	8.4
Assistant to senior management			*Assistant to senior management*		
Assistant chief	11	9.3	Deputy manager	38	4.4
Assistant manager	1	0.8	Group leader	3	0.3
			Assistant group leader	3	0.3
			Assistant manager	82	9.5
			Assistant specialist manager	43	5.0

			Head of laboratory	4	0.5
			Head of overseas office	9	1.0
			Assistant head of laboratory	6	0.7
			Assistant chief of laboratory	1	0.1
			Assistant manager in charge	1	0.1
Sub-total	12	10.1		190	21.9
Middle management			*Middle management*		
Section chief	55	46.2	Group leader	6	0.7
Room chief	5	4.2	Head of business office	16	1.8
			Head of steel yard	1	0.1
			Chief doctor	1	0.1
			Room chief	31	3.6
			Section chief	487	56.3
Sub-total	60	50.4		542	62.6
Total	119	100.0		860	100.0

Sources: Tekkō Shimbun-sha 1960; Daiyamondo Publishers 1979

to manage this problem, companies tended to increase the number of staff functions in senior management, assistant to senior management and middle management classes. Those who fail to become core managers thereby need not be discouraged in the so-called life-time employment system. We can observe this phenomenon across the corporation but most typically at head office. Partly it is also due to an increase in and division of specialised management functions. Middle management expanded principally due to an increase in percentage of section chiefs. The expansion, division and specialisation of managerial positions below senior management class indicates the relative decline in the status of managerial personnel and increased competition for promotion among them. In terms of content of daily jobs the percentage of attendance at meetings and producing and compiling data and materials has increased. Office automation has transformed the nature of middle management work into standardised administrative tasks more than managerial ones.[7]

In comparison, we can observe that promotion to director became more difficult in 1978 compared to 1960. In 1960 it was possible for the majority of senior management employees to be promoted to such positions, while in 1978 it became very difficult in terms of the gap in numbers between the director and senior management. The same can be said of promotion to senior management from middle management. In addition, we may conclude that the increase in the assistant to senior management level is due to a combination of two factors: an increase in those who failed to enter senior management and an increase in candidates for senior management. Both categories of people are qualified for senior management in terms of official qualifications; however, one exists because of people who failed to be promoted, while the other consists of those just waiting for further promotion. The *Nōryokushugi Kanri*, which has been explained earlier, has achieved a compromise to solve the problem of a shortage of posts for potential managers, which was caused by collective seniority promotion in long-term employment. The increase of the assistant to senior management class is a reflection of this compromise.

In reality, whether one can be promoted or not to core manager in each management class depends upon three factors: ability to carry out the job; personality; and potential contribution to the organisation. Ability to do the work required accounts for 50–60 per cent, personality – which includes whether one is favoured by one's immediate superiors or not – accounts for 30–40 per cent, and the potential contribution element accounts for 10–20 per cent. In qualitative terms, a transfer which helps to foster human relations with influential department managers can be understood as a good promotion, while a 'negative' kind of promotion also exists.[8] For head office employees who have not arrived at senior management level, such a transfer occurs almost every four years and is usually a move between head office and steel works.

Apart from official qualifications for and process of promotion, one needs to have a good relationship at each level with influential managers whose line of connection reaches as high as possible in the hierarchy. After about five years from entering the company one will begin to know one's relative place and position and be able to make a forecast as regards the area and level of future promotion.

EDUCATION AND TRAINING OF CLERICAL, TECHNICAL AND MANAGERIAL STAFF

Investment in human resources implies a capacity for returns which exceed the original investment. In management practice, it is done because rapid organisation growth necessitates it. The growing numbers of white-collar staff require systematic management of those employees, and in-firm education and training provides it appropriately, while enhancing employee skills, knowledge and company loyalty. Company statements of their educational philosophy do not always accord with their subsequent education and training programmes, in the sense that they represent the aspirations and ideals of the Faculty Development Department or other department set up for the purpose. Yet the various statements and programmes to be introduced and examined here can be seen as an early stage of human resource approach in the large Japanese corporations, where 'life-time' employment is practised and industrial relations are 'harmonious'.[9] They are the reflections of management ideals concerning how human resources are to be nurtured for the growth of the enterprise and good industrial relations, and stand in marked contrast to any views expressed by the BSC management.

Responding to the necessity of integrating workers into the company's needs, the Japan Iron and Steel Federation in 1962 published 'Education Systems and the Collection of Education Regulations', a report which outlines the approach of each member company and describes the principles of their employee education. Yawata Steel Corporation's approach is described thus:

> The purpose of employee education is to develop the temperament and ability of employees needed for business, let them recognise the

> importance of the mission given to the company, let them co-operate with each other as members of the organic enterprise entity, and as a result to execute the goals of business efficiently. Therefore, employee education is an education contributing to business development, and not a general and cultural education. The distinction between the two is not always clear, but in any case it must be one required by the necessity of the business and it must contribute to the prosperity of the business.
>
> (Tekkō Renmei 1962: 3)

And for Kawasaki Steel Corporation:

> To survive severe competition and maintain as well as develop the existence of the enterprise, in the midst of technological innovation, the accomplishments of the human elements, the employee, which is the fundamental power needed to undertake the above task, is vitally important. The purpose of school education is to provide the learning necessary for a member of society, i.e. education for the development of personality, while that of the enterprise is to pursue the most efficient production (including clerical service work). As a matter of course, the education of the enterprise includes the development of personality as an important component, but it is supposed to be pursued persistently 'through production', and accomplishments as an employee will be attained through the education and training given after one enters the company.
>
> (Tekkō Renmei 1962: 37)

The common recognition of the purpose of education of these companies is that it contributes to production and business, and is not simply for the personal development of the employee. They both, like their rivals, face severe competition and are in the process of rapid technological innovation. In this context, education and training specific to the objectives of the company are required. The education and training within the enterprise is therefore distinguished from general education. In other words, the emphasis is on education to increase efficiency and profitability, and in this sense it becomes an important part of management. And all this trend in education and training reflects that the companies are on an upward slope of development, namely at the stage of 'positive' compatibility. A long-term and systematic in-firm education and training was necessary as a catalyst to create institutional arrangements for employment management capable of fostering higher compatibility in management.

Tables 7.6 and 7.7 show the characteristics of the typical education

Table 7.6 Plan of education for university graduates in a large steel company (Yawata Steel Corporation)

Type of education	*Period of education*	*Content of education*	*Method of education*
Introductory education	1 month	To provide knowledge of organisation and business of the company and give a work attitude as members of the company	Sit-in lectures by Education Deptartment, providing work opportunities in factories
Basic education	About 2 years	General management techniques and fundamental knowledge of the workplace, as applicable to any type of staff work	1 Clerks 1st year: Attachment to a job in factory and provision of workplace education based on the basic education plan 2nd year: Attachment to a job in an office and provision of education based on the basic education plan
Education by classification	From after basic education until becoming chief clerk	Based on the classification of staff (A type, B type, C type), to provide necessary job experience and special management techniques and professional knowledge	2 Engineers Attachment to Staff Deptartments and provision of education based on the basic education plan (a) Job change according to plan (b) Attendance at a class and outside institute

Source: Tekkō Renmei 1962: 12

Table 7.7 Plan of education for high school graduates in a large steel company (Yawata Steel Corporation)

Type of education	*Period of education*	*Content of education*	*Method of education*
Introductory education	3 Weeks	Similar to that of university graduates	Similar to that of university graduates
Supplementary education for engineers	Basic knowledge 37 days Professional knowledge 5 months	Provide the graduates of Technical High School with professional knowledge necessary for daily work	Supplementary evening classes by Education Department
Education by classification	Same as university graduates	As university graduates	As university graduates

Source: Tekkō Renmei 1962: 13

for newly recruited university and high school graduates in the early 1960s. Differences in the period and content of education exist according to the school career of the employee. While university graduates have one month's introductory education and then about two years' basic education, high school graduates have only three weeks' introductory education and no further basic education.

The basic education for university graduates is divided into two programmes, one for clerical and the other for technical employees. For high school graduate technical employees, there are supplementary lectures to supply basic and professional knowledge. The former is for thirty-seven days and the latter for five months. As shown in Table 7.6, the specialised education is described as being the same for university and high school graduates, although in reality the content, time and period is thought to have been different, as university graduates underwent it after two years' basic education and high school graduate clerical employees had it after three weeks' introductory education. Although there is no data for female employees of Yawata Steel Corporation, education for women at Kawasaki Steel Corporation was very short, consisting of only ten days' comprehensive education.

We have reports on managerial staff education for Yawata and Kawasaki Corporations:

Yawata Steel Corporation:

> Education for managerial staff is usually done by having them attend seminars and study meetings sponsored by outside institutions. These seminars and meetings extend over several days during the daytime. There is also training given in the enterprise during working hours such as MTP (Management Training Program) and IE (Industrial Engineering).
>
> (Tekkō Renmei 1962: 13)

Kawasaki Steel Corporation:

> Training for top management staff is given by having them attend various lectures and meetings held both inside and outside the company. This aims to have them acquire a broad knowledge of business management and promote self-development. In addition to this, we prepare various pamphlets and materials introducing the education of other corporations, hoping that they will recognise the importance of education and promote it positively in the company. The emphasis in the training of managerial staff (managers and section chiefs) is on enhancement of understanding of the modern business management system: encouragement of self-development, methods of management, improvement of work methods, etc. This is achieved by having them attend various lectures and seminars held at various places including inside the company and by distribution of reference materials on business management.
>
> (Tekkō Renmei 1962: 42–3)

The principle introduced here shows that although the necessity for education of managerial staff was recognised, it was not yet established and practised systematically in the enterprise. The main method was to encourage members of the board and managerial staff willingly to attend lectures, seminars and study meetings organised by various institutions. In the mid-1960s, reflecting the increase in size and organisational development of the steel companies, which was caused by the establishment of new large-scale works, and also stimulated by severe domestic and foreign competition, the large steel companies recognised the significance of education as an important part of human resource management for their growing and diversifying managerial staff. The following are the basic outlines of managerial education proposed by representative steel companies in this period (Tekkō Renmei 1966: 113).

Yawata Steel Corporation:

> With the liberalisation of trade as a momentum, conditions for the survival of Japanese enterprises have become more difficult, and

competitive pressure has become even stronger. Under such a situation, the managerial staff in the new age should not only have professional and technical knowledge and skill, but also must be co-operative with various people in the team, thus managerial staff who can comprehensively apply varied, complex and high professional knowledge and skill must be created.

Also, managerial staff should make efforts to provide appropriate information so as not to mislead the top management who are in charge of an extremely large enterprise with diversified functions in the long view. Enough knowledge, insight and power to act should be given systematically with good planning through education in order to meet the above mentioned requirements.

(Tekkō Renmei 1966: 113)

Fuji Steel Corporation:

1 Put emphasis on fostering far-sighted vision, and learning the fundamental principles of job management and special needs of the assigned job, thus achieving the development of leadership necessary for the managerial personnel.
2 The purpose of education will be achieved by self-edification, daily individual guidance and collective education.

(Tekkō Renmei 966: 113)

Nippon Kōkan, KK:

Due to the recent remarkable progress of technology and the development of the steel industry, competition in the international market has become further intensified. We are moreover urged to prepare measures to create excellent managerial personnel.

In order to create managerial personnel who can cope sufficiently with new situations under such severe circumstances, the emphasis of managerial education is put on improving basic knowledge, techniques and skills of business management, thus enhancing far-sighted vision and leadership to a higher level. Desire for self-improvement is a prerequisite; collective education will be given whenever necessary as well as giving opportunities to get education at institutions other than the company.

(Tekkō Renmei 1966: 113)

Kawasaki Steel Corporation:

1 Achieving a common understanding of the measures for the practice of basic managerial policy of each department and business ideology of the company.

2 Provision for and development of the variety of knowledge and skills of scientific management necessary for modern business administration.
3 Provision for and development of knowledge and skill for guiding and training subordinates with an aim of preparing successors.
4 Giving motivation for self-improvement and developing the far-sighted vision and personality needed as managerial staff of the enterprise.

(Tekkō Renmei 1966: 113)

Sumitomo Metal Industries Ltd:

> Realisation of the long-term equipment development program, the development of managerial structure due to the expansion of business, the development and increase in complexity of technology, the introduction of the line and staff system into all establishments since 1961, and other changes inside as well as outside the enterprise have caused it to be necessary to unify the managerial ideas of the managerial personnel and develop management ability and judgement.
>
> (Tekkō Renmei 1966: 113)

The reasons and principles stated by these companies are more or less the same. Depending upon the corporate environment occupied and perceived by each company, slight differences appear in their emphasis. Yawata Steel Corporation, which was at this time the largest steel company in Japan, sensed the threat of overseas competition most and attempted to develop management education and training with that situation in mind. Late developers such as Kawasaki and Sumitomo instead directed their attention more to the importance of the management skills and knowledge immediately required by their recent organisational growth.

Thus, managerial education was necessitated by the growth of the company and the intensification of competition, and among the abilities required for managerial personnel were professional managerial knowledge and techniques, human relations and leadership. The method of education is both collective, with various seminars held by institutions exterior to the companies themselves, and by individual guidance. It is interesting to see some cultural tradition of Confucianism in the form of wording which suggests self-improvement and self-edification. It is also mentioned that managerial function has increased and has become diversified as well as specialised due to the expansion of business made possible by plant and equipment modernisation. We have examined this point in connection with the development of various organisations such as labour management-related departments, general research departments, sales management departments, export-related departments, technology services departments, production control departments, equipment technology centres, R & D departments, and planning and development departments of the head office, and production departments, technology departments and equipment departments of the works, in the case of the Nippon Steel Corporation. The emphasis of education was in particular put on the section chief positions, which increased both in number and percentage (Tekkō Renmei 1972: 75).

Figure 7.1 shows a managerial education system in operation at a major steel corporation in the mid-1960s. This consists of a hierarchically-aligned study and training system for chief clerks, section chiefs and managers; and special lectures for the various departments and professional divisions. Management Training Programme (MTP) lectures, which newly appointed chief clerks will attend, and a study and training programme for chief clerks aims at giving them concrete and professional managerial knowledge and the skills necessary to become first line managerial personnel. Section chiefs are in charge of the above mentioned managerial education. The study and training for section chiefs aims at teaching them methods of scientific management, scientific thinking in decision-making, professional knowledge and higher level skills to solve various managerial

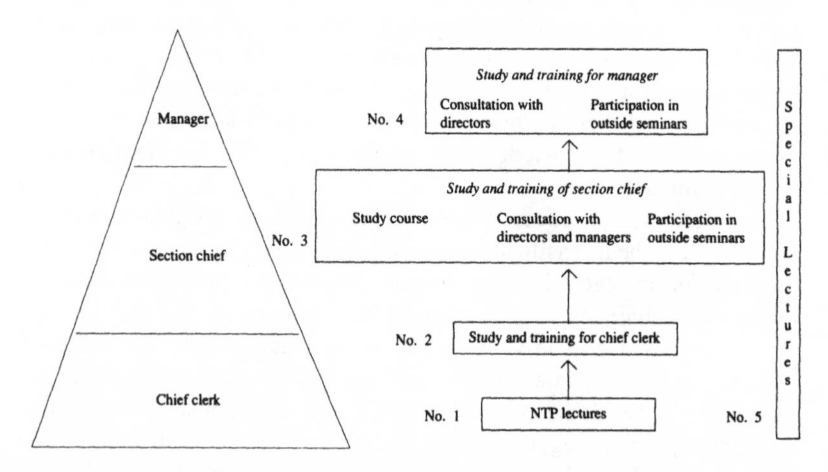

Figure 7.1 Education system of managerial staff in large-scale iron and steel enterprise (Yawata Steel Corporation)
Source: Tekkō Renmei 1966: 1

problems; and giving them a broad vision and knowledge about business management through various meetings with managers and directors as well as attending seminars held outside the company. Lecturers for such programmes are general manager-class personnel. The study and training for managers is mostly done through various meetings attended by directors and seminars held outside the company. The object of education for managers is to improve the ability of senior management in planning, organising, controlling, and supervising the division of which they are in charge. Special lectures were given for the various departments and professional divisions, while industrial engineering study and training meetings were held for managerial personnel in the maintenance department.

The emphasis of managerial education is put on the hierarchy of study and training. At higher managerial positions, the character of education changes from learning concrete managerial knowledge and skills for the execution of managerial work, to acquiring the knowledge and ability suitable for planning, organising, controlling and supervising the work necessary for senior management. This development and systematisation of managerial education suggests that division of managerial functions within the enterprise has developed to the point of diversification and specialisation of managerial personnel. Thus all managerial personnel are required to contribute efficiently to the objectives of the company under the hierarchical personnel management structure.

By the beginning of the 1970s, employee education for clerical, technical and managerial personnel had become a part of the *Nōryokushugi Kanri* of the whole company. This is a Japanese version of human resource management which accommodates the existing education and training scheme into the basic structure of personnel management across the corporation. One of the objectives is to modify the nature of traditional seniority-based promotion to that of individual assessment-based promotion. Although the percentage of this individual assessment to the whole salary/wage was initially low, it gradually increased to 50 per cent of the whole salary/wage in the late 1980s. Education and training was used as an important tool to qualify employees for a position of qualification determining the above portion of salary/wage payment. At the same time, it solved the problem of imbalance between the number of line posts and the number of possible candidates for such posts, as discussed above. A number of employees had to remain qualified for certain line posts in senior management but unpromoted, yet still this system helped greatly to reduce dissatisfaction among those who failed to obtain the line posts as well as to keep up the morale of the employees. Thus it was necessitated by the emerging negative effects of seniority promotion which began to appear as organisational growth drew nearer to termination in the early 1970s.

Figure 7.2 shows the structure of a typical *Nōryokushugi Kanri* of a representative large steel company. Education for clerical and managerial personnel – which is half of the whole system, and which includes blue-collar employees – will be examined here. Employee education at this time was linked to the *Shokunō Shikaku Seido* (Job Qualification System) which took on complete form in the early 1970s. The qualification titles are shown in the centre of Figure 7.2, and represent level of education as well as corresponding line positions; qualifications including part of and above the *Sanji-ho* level represent managerial positions. There are three kinds of programme: study and training at the workplace; professional study and training; and hierarchical study and training. The former two programmes are separate from qualification and position, but the hierarchical programme will be given to those who achieve one step higher qualifications. For example, a manager obtaining a *Sanji-ho* qualification will receive the hierarchical education required for group manager. As for qualifications below and including *Shuji*, at the time of entering the company high school graduates and two-year women's college graduates will be given a qualification of *Tantō-ho*; technical college graduates, *Tantō*; university graduates, *Shutantō*; and graduates of a postgraduate course, *Shuji*. Employees with a qualification of *Tantō-ho*, *Tantō* and *Shutantō* will have an introductory study and training programme as shown in Table 7.8. The time period of the programmes differs depending on school career and the establishment where they start work. The higher the school career years, the longer the programme; the head office has a longer programme than the steel works, the new steel works has a shorter programme than the old steel works, the female high school graduates have a shorter programme than male high school graduates. Iron and Steel Technical College graduates are not attached to the head office, but to the steel works. The employee with a qualification of *Tōkatsu Shuji*, which is between ordinary and managerial personnel, will attend a study and training programme entitled the 'Leader Programme', under which they study with an individual theme. Those who have a *Sanji-ho* qualification or above will receive managerial education, given at manager and group manager study and training meetings, and managers' social meetings.

As is made clear in the above, the hierarchy of the study and training programme linked to the *Shokunō Shikaku Seido*[10] as part of *Nōryokushugi Kanri* was the central part of the employee education for

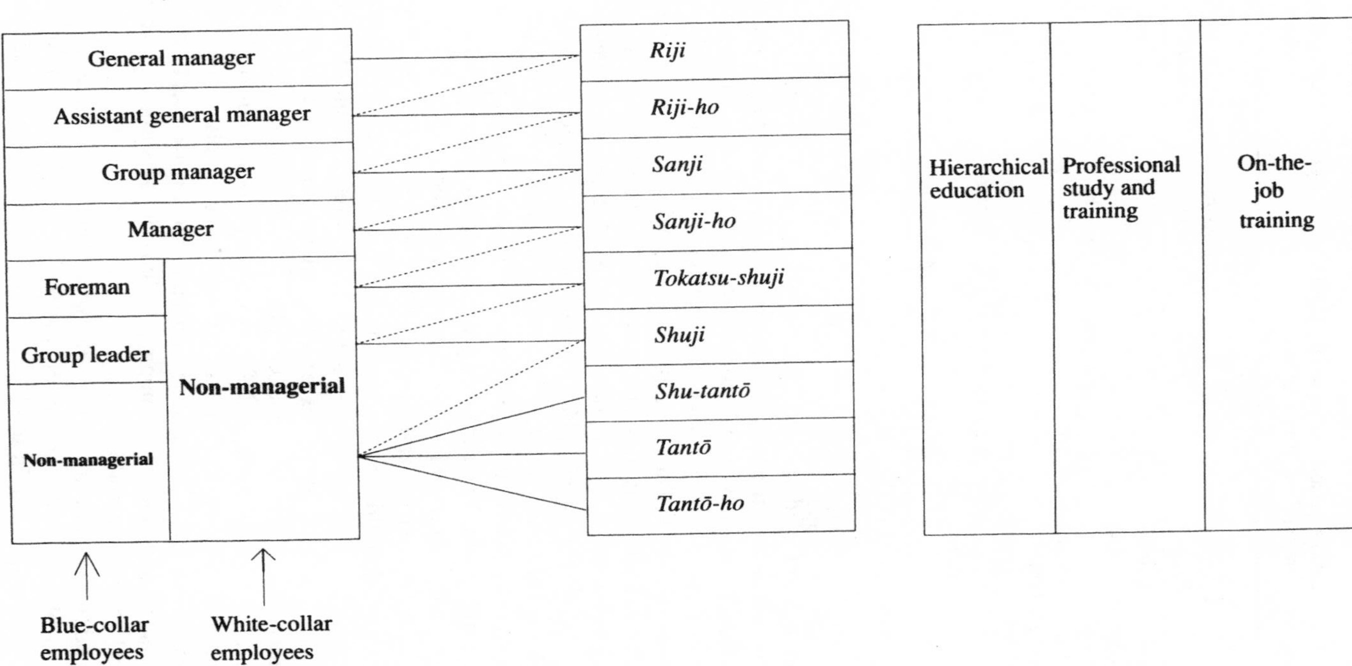

Figure 7. 2 System of *Nōryokushugi Kanri* (Nippon Steel Corporation)
Source: Tekkō Renmei 1977: Nippon Steel, Interview

Table 7.8 Education period for new employees of a large steel company by school career (Nippon Steel Corporation)

	Head office	*Old steel works (Yawata)*	*New steel works (Kimitsu)*
University graduates and technical college graduates	3 nights 4 days	3 months	76 days
Iron and steel junior college graduates		2 days	2 days
High school graduates (male)	3 nights 8 days	2.5 months	2 days
High school graduates (female)	11 days	8 days	

Source: Tekkō Renmei 1977

clerical, technical and managerial personnel in the 1970s and 1980s. In addition to this hierarchical programme, there was on-the-job training and professional study which contributed to the technical aspect of education and training. School career and gender have much to do with employee education. This implies that the division of work in clerical and technical labour is similarly based upon school career, sex, and employee education. The school and gender differences in education terms existed in the early 1960s, and in the 1970s the period of study and training for new employees became greatly reduced, except for the 2½-month programme for male high school graduate employees attached to the old works. The period of programmes for learning individual and technical managerial skills has also been reduced in the 1970s compared to that of the early 1960s (Tekkō Renmei 1977: 5–26). We may assume that this contraction of the time period for employee education was made possible by the successful recruitment of various workforces, such as graduates of postgraduate courses, universities, iron and steel colleges, technical colleges (all men only), and women's colleges, as well as male and female high school graduates. Concomitant with the diversification of the workforce, employee education was also diversified and specialised, thus providing more vocational and rationalised education based upon the division of work in clerical, technical and managerial labour.

Education and training as part of overall human resource management has played a very important role in management, in particular personnel management, in solving the emerging problems of traditional seniority promotion, by both reducing the labour costs and boosting declining morale.

CONCLUSION

Management systems are in constant flux to adapt to the economic situations affecting their company and industry. This chapter has examined the changes that occurred at head office level in the steel companies, looking at their expansion, division of labour and development of in-firm education and training systems. These developments reflected the quantitative growth enjoyed by the steel companies and when that growth came to an end, adjustments were made, such as in the length of education and training. On the whole, formal means of human resource management served well to maintain control and integration in the growing organisations, but these were usefully underpinned by informal patterns of human relations. The general impression of the circumstances of management and labour at head office must therefore be one of organic compatibility and a gradual shift, through *Nōryokushugi Kanri*, towards contrived compatibility.

Part III

The British steel industry

8 The British steel industry

Modernisation in the British steel industry proceeded differently from that in Japan, in that the timing of investment and the scale and content of modernisation were mainly caused by political conditions and by the lower levels of demand in the declining British manufacturing industry. The similarities in the form of modernisation can be understood as due to a diffusion effect of a common production paradigm. Modernisation and rationalisation have caused unique changes in management and labour, based upon the already-established division of labour and social relations in the British steel industry. It reflects a stage of contrived compatibility and the consequences of modernisation and rationalisation suggest a potential for some convergence towards Japan in the use of the workforce at the workplace.

PLANT AND EQUIPMENT MODERNISATION

Throughout the 1960s the growth rate of the UK economy was relatively low, stultifying the modernisation and rationalisation of the iron and steel industry. The Labour government elected in 1964 effected the renationalisation of the major steel companies in July 1967 under the Iron and Steel Act of 1967. In the 1970s plant and equipment modernisation and rationalisation were promoted extensively by the British Steel Corporation (UK ISSB, *Annual Statistics*: 1960–80).

Actual plant and equipment investment in the steel industry in relation to that of the whole manufacturing industry increased from 7.9 per cent in the period 1962–9 to 11.6 per cent in 1970–8. In fact, the total amount of investment in the 1960s was £1,222 million, while in the 1970s it was £4,177 million, a 340 per cent growth. The most marked increase occurred during 1960–2; in these three years, 46.6 per cent of the total investment for the 1960s was spent, setting the main trends and content of the plant and equipment modernisation for the decade. These were manifested in the increase of investment in steelmaking and rolling, with related equipment modernisation, and the decrease in coke production, raw materials, ironmaking and related areas. For example, between 1960 and 1965 the percentage share of investment in coke production decreased from 5.8 to 1.1 per cent, and ironmaking and related equipment from 20.6 to 15.6 per cent, while investment in steelmaking increased from 12.7 to 17.2 per cent and rolling and finishing from 57.3 to 58.3 per cent (Iron and Steel Board 1960, 1965). In the second half of the 1960s, investment was lower than the first half, influenced by nationalisation, and there was no discernible change in the trends of investment by production processes.

The above trends in plant and equipment investment in the early 1960s are indicated by installations of LD converters, continuous casters, hot strip mills and cold strip mills (Rowley 1971: 112–14). The representative corporations and works in which plant and equipment modernisation was promoted during 1960–2, i.e. the peak of 1960s investment, are listed below (Iron and Steel Board 1960, 1961, 1962):

1 Appleby–Frodingham Works (The United Steel Companies Ltd)
2 Ebbw Vale Works, Spencer Works (Richard Thomas & Baldwins Ltd)
3 Margam Works, New Port Works (The Steel Company of Wales Ltd)
4 Ravenscraig Works (Colvilles Ltd)
5 Corby Works (Stewarts and Lloyds, Ltd)
6 Lackenby Works (Dorman Long – Steel – Ltd)
7 Guest Keen Iron and Steel Works, Scunthorpe Works (GKN Steel Co. Ltd)
8 Tinsley Park Works (English Steel Corporation Ltd)
9 Rotherham Works (The Park Gate Iron and Steel Co. Ltd).

Plant and equipment investment in the 1970s was much larger in the second half of the decade than the first, and was especially concentrated during the three years from 1975–7, accounting for 43.7 per cent of the total investment of the 1970s. BSC's share of the total investment was extremely high, at 93.3 per cent at its peak in 1974–6, but subsequently showing a marked decline down to 62.5 per cent in 1980.[1] During the period of the highest investment (1975–7), the Corporation invested £1,585 million, of which 38.6 per cent was in raw material and related equipment investment, 18.6 per cent in ironmaking and related equipment, 20.6 per cent in steelmaking and related equipment and 22.3 per cent in rolling and finishing equipment. As distinct from the plant and equipment investment in the first half of the 1960s, in the 1970s attention shifted to promoting the raw material, ironmaking and related areas. In addition, during 1975–7 £230 million was spent on each of the Ravenscraig Stage 3 and the Redcar Phase 2-B developments. Works representative of large-scale plant and equipment modernisation are the Ravenscraig, Scunthorpe, South Teesside and Port Talbot Works, all of which were major steelworks prior to nationalisation.

MODERNISATION BY PRODUCTION PROCESS

Plant and equipment modernisation after 1960 in the major advanced countries was represented by large-scale, continuous, high-speed and automated equipment introduced into production processes. The outline of the British plant and equipment modernisation was (as shown in Table 8.1) to enlarge blast furnaces, do away with open hearth furnaces and install LD converters, continuous casters, hot strip mills and cold strip mills. In the 1960s LD converters and strip mills increased in number, as did continuous casters in the first half of the 1970s. During this decade, the use of small-scale blast furnaces and open hearth furnaces was largely discontinued. The failure of steelmaking capacity and crude steel output to grow any further beyond the peak year of 1970 testifies to the stagnation and in some cases decline of the steel-consuming industries, and between 1960 and 1980 employment in the industry fell dramatically. In these twenty years, 171,900 jobs were lost, especially in the latter half of the 1970s.

Ironmaking process

Equipment modernisation here consisted of scrapping old small blast furnaces and replacing them with larger ones. In 1960, eighty-nine blast furnaces with a hearth diameter of under 20 feet (inner volume 640 m^3) accounted for 80.9 per cent of the British total, but by 1980 there were only five left, representing 21.7 per cent of the total. The largest blast furnaces in 1960 had a hearth diameter of 31 feet (inner volume about 1,874 m^3) and were located at the Appleby–Frodingham Steel Co. (two furnaces) and the Steel Co. of Wales Ltd (one furnace). After that, blast furnaces of over 28 feet hearth diameter increased to twelve in 1970, seventeen in 1975 and twenty in 1978. In 1980 six blast furnaces exceeded 30 feet; the largest of all, at 45.9 feet (inner volume 3,900 m^3), commenced operation at Redcar Works in October 1979. These six were:

1 South Teesside – Redcar works (one blast furnace, hearth diameter 45.9 feet)

Table 8.1 Modernisation of the British steel industry

	Investment expenditure (£ million)	*Crude steel output (1,000 tonnes)*	*Steel making capacity (million tonnes)*	*Units of equipment*					*Number of employees (iron and steel, general) (thousand)*
				Blast furnaces in use	*Steel furnace (LD)*	*Continuous caster*	*Hot strip mill*	*Cold strip mill*	
1960	606	24,695	25.8	85 (188.4)	743(33.2) [1]	2	3	11	326.1
1965	763	27,421	32.0	66 (267.1)	739(37.1) [15]	n.a.	6	19	305.0
1970	459	28,291	33.4	56 (316.0)	361(78.3) [18]	11	6	20	281.5
1975	1,210	20,098	27.0	35 (346.6)	262(76.7) [20]	23*	6	25	240.0
1980	2,967	11,277	28.4	9 (665.5)	112(100.7) [16]	28	6	25	154.2
1985	1,515	15,722	n.a.	11 (950.8)	53(296.6) [14]	25	n.a.	n.a.	84.0

Sources: UK Iron and Steel Statistics Bureau 1960, 1967, 1970, 1975, 1978, 1980, 1985; CSO *Annual Abstract of Statistics* 1968, 71; The Iron and Steel Board 1960, 1964; Industry and Trade Committee (Session 1980–81, Volume 11, *Effects of BSC's Corporate Plan*: 229; Tekkō Renmei, Tōkei Yōran 1961, 1966, 1975, 1976, 1980; Tekkō Kyōkai 1976: 7

Notes: (a) Investment expenditure indicates the total amount in the preceding five years
(b) Crude steel output indicates the output of the year shown
(c) ( ) output per furnace in 1,000 tonnes. [ ] the number of LD converters
(d) * the number of continuous casters in 1974

2 Appleby–Frodingham works (one blast furnace, hearth diameter 31.0 feet)
3 Llanwern works (one blast furnace, hearth diameter 36.7 feet and two blast furnaces, hearth diameter 30.9 feet)
4 Port Talbot works (one blast furnace, hearth diameter 31.0 feet).

(UK ISSB 1980)

The above modernisation of the ironmaking process resulted in a 353.2 per cent increase in output per furnace, from 188,400 tonnes to 665,500 tonnes, between 1960 and 1980 (see Table 8.1).

Steelmaking process

In this area, open hearth furnaces were replaced by LD converters and continuous casters with enhanced capacity. The 350 open hearth furnaces in 1960 accounted for 84.5 per cent of the total crude steel production, but after that numbers decreased to 157 in 1970 and eighty-three in 1975, until in 1980 the use of all open hearth furnaces was discontinued. The first LD converter was installed in 1960 at the Ebbw Vale works of Richard Thomas & Baldwins Ltd. The number increased to fifteen in 1965, eighteen in 1970 and twenty in 1975, but in 1980 dropped to sixteen. Concurrently, LD converters increased in size; the largest in 1965 took 128 tonnes per charge, the largest in 1970 300 tonnes, 1975 328 tonnes and 1980 330 tonnes. The percentage share of LD converters in total crude steel output increased from 0.1 per cent in 1960 to 32.1 per cent in 1970 and 59.3 per cent in 1980 (Iron and Steel Board 1964: 38; UK ISSB 1980).

In 1980 sixteen LD converters were installed in the following major BSC works:

1 South Teesside – Lackenby works (3 x 260 tonnes)
2 Appleby–Frodingham works (3 x 300 tonnes)
3 Normanby Park works (2 x 86 tonnes)
4 Llanwern works (3 x 175 tonnes)
5 Port Talbot works (2 x 330 tonnes)
6 Ravenscraig works (3 x 130 tonnes).

The equipment modernisation of the steelmaking process did not lead to higher steelmaking capacity overall, but increased the output per furnace from 33,000 tonnes to 100,000 tonnes between 1960 and 1980 (see Table 8.1).

From two continuous casters in 1960, numbers rose to eleven in 1970, twenty-three in 1974 and twenty-eight in 1980. Hence, the percentage of continuously-cast steel went from 0.4 per cent of the whole output in 1963 to 1.3 per cent in 1965, 1.7 per cent in 1970, 8.5 per cent in 1975 and then a steep rise to 27.1 per cent in 1980. The number of casters increased rapidly in the first half of the 1970s and during the latter half of the decade the percentage of continuously-cast steel rose remarkably. In 1980 there were twelve casters in small- and medium-sized private producers, of which there were three slab casters, one bloom caster and eight billet casters. The Corporation owns nine slab casters, five bloom casters and two billet casters in the following major works:

1 South Teesside – Lackenby works (one bloom caster, two slab casters)
2 Appleby–Frodingham works (two slab casters)
3 Ravenscraig works (three slab casters).

Rolling process

The equipment modernisation in this process is seen in the installation of hot strip mills and cold strip mills. In the first half of the 1960s the number of both increased rapidly, but after 1965 hot strip mills remained static. The number of cold strip mills began to increase again in the first half of the 1970s (see Table 8.1). The number of cold strip mills in 1980 was twenty-five, including twelve tandem mills, six reversing mills and seven sendzimir mills. Hot strip mills in 1980 numbered three full-continuous mills, two semi-continuous mills and one steckel mill, installed in the following five works:

1 Llanwern works (3,035,000 tonnes p.a.)
2 Port Talbot works (2,200,000 tonnes p.a.)
3 Ravenscraig works (1,970,000 tonnes p.a.)
4 Rotherham works (400,000 tonnes p.a.)
5 Corby works (two mills, 985,000 tonnes p.a.).

(Metal Bulletin Books 1978)

We can see therefore that, in summary, modernisation consisted of replacing older blast furnaces with new large ones, and siting those at the major BSC works. At the same time, LD converters and continuous casters were brought in to do the work of open hearth furnaces and, by their greater size, enhance steel production; the LD converters are located in the six major works of the Corporation, sixteen continuous casters are located in the three major works of the Corporation and another twelve continuous casters are in the works of private producers.

In the rolling process, hot strip mills and cold strip mills were installed and the six hot strip mills are located in the five major works of the Corporation. The most modernised integrated works in terms of iron and steelmaking (inclusive of continuous casters) are Redcar/Lackenby (main products: sections, hot rolled coil, billets, bars and rods, piling and colliery arches) and Appleby–Frodingham (main products: sections, plate, billets, bars and rods). Llanwern (main products: hot and cold rolled coil and sheet) and Port Talbot (main products: hot and cold rolled, and galvanised, coil and sheet) are also modernised in the iron and steelmaking and rolling processes, but as of 1980 continuous casters had not been installed.

MODERNISATION OF THE BRITISH STEEL CORPORATION

The plant and equipment modernisation of the UK iron and steel industry since 1967 is represented by the reorganisation of the Corporation's steelworks; the scrapping of old and small works and factories, and the concentration of production in enlarged and modernised existing large-scale works.

The Corporation, when established in 1967, had twenty-one integrated works, eighteen steelmaking–rolling works, nine rolling and other works, 153 subsidiaries, nine related companies and forty-eight overseas subsidiaries and related companies (BSC 1967). Thereafter, rationalisation took these figures by 1980 to six integrated works, eight steelmaking and rolling works, fifteen rolling and other works, twenty-three subsidiaries, forty-seven related companies and forty-seven overseas subsidiaries and related companies.[2] Thus, there has been a large decrease in the number of integrated works, steel and rolling works and subsidiaries and an increase in the number of rolling and other works and related companies. This last increase was caused by scrapping the iron and steelmaking processes of the older and smaller integrated works and also the steelmaking processes in the old and small steel and rolling works. The increase of related companies was due to the sale of shares in the wholly-owned subsidiaries of the Corporation. The fact that there was no change in the number of overseas subsidiaries and related companies suggests that overseas business activities were healthy, in contrast to the domestic stagnation of the Corporation.[3] Thus the Corporation was and remains an enterprise which owns and controls a large number of subsidiaries and related companies in various business areas both at home and abroad.

In size, iron and steel works in 1966 consisted of one site with a capacity range of steel production between 3–4 million tonnes, nine between 1–2 million tonnes, seven between 0.5–1 million tonnes and fifteen under 0.5 million tonnes. Of the total UK steel output, more than

90 per cent of steel was produced by these works (Hunter *et al.* 1970: 118). Among these thirty-two works, there were twenty-one integrated iron and steel works. In 1978 there were seven integrated works, of which five had a capacity of over 3 million tonnes, and the Scunthorpe Works had a capacity of 5.9 million tonnes.

Plant and equipment modernisation

The progress of plant and equipment modernisation and rationalisation of the Corporation in the 1970s was based upon the *Ten Year Development Strategy* (February 1973),[4] the *Road to Viability* (March 1978),[5] *The 15 Million Tonnes Strategy* (December 1979)[6] and the *Survival Plan* (December 1980).[7] In the *Ten Year Development Strategy*, a massive amount of plant and equipment modernisation was planned, with the recognition that the modernisation and expansion of the steel industry was vital, it being a key industry for the development of the UK economy. The total amount of plant and equipment investment was to be £3,000 million over the next ten years. This plan was based on the assumptions that the demand for steel in the 1970s would increase; international competition would intensify; large-scale coastal works (works with crude steel production over 6 million tonnes) were most favourable in terms of economies of scale; and that UK equipment investment in the second half of the 1960s had been small compared with other major countries and many old low-productivity works had been retained. The majority of investment was to finance the modernisation and expansion of Port Talbot, Llanwern, Scunthorpe, Lackenby and Ravenscraig works to their optimum size. The capacity for crude steel production was planned to increase to 33–35 million tonnes towards the end of the 1970s, and to 36–38 million tonnes in the early half of the 1980s. During this period the use of open hearth furnaces would be discontinued and many works and factories would gradually be rationalised. Between 1967 and 1972, staffing was to be reduced by 31,450, and by a further 50,000 in the next ten years. New employment measures were considered in areas where the number of employees was reduced and various projects planned to assist those who chose transfer, retraining, or early retirement.

The *Road to Viability* aimed to realise early closures of low-productivity works and factories, and reduce the plant and equipment investment for large projects as, contrary to the forecast of the *Ten Year Development Strategy*, the demand for steel products fell to its lowest level in forty years. With regard to large projects, plans for the hot strip mill at Port Talbot and plate mill and tin plate developments were shelved. The Clyde Iron Works, the Hartlepool Steel Works and East Moors Steel Works had already been slated for closure and negotiations to close Ebbw Vale had started. Various measures were planned to create alternative employment to cushion this employment rationalisation and it was aimed to raise manpower productivity to the level of leading European countries, the USA and Japan.

The *15 Million Tonnes Strategy* planned to reduce the capacity for crude steel production of 19.8 million tonnes at the end of 1979 to 15.2 million tonnes by January 1981. This reduction could be achieved by closing works at Consett and Hallside, heavily reducing production capacity at Scunthorpe, Llanwern and Port Talbot and increasing it at Lackenby and Ravenscraig. The number of employees was to be cut from 186,000 to 143,300. Table 8.2 shows the re-organisation plan for the reduction in steel production capacity and employment by the steel works concerned.

The Survival Plan recommended reduction in crude steel production capacity to 14.4 million tonnes in the face of a large deficit and stagnation of demand, and the increase of capacity utilisation from 64

Table 8.2 Re-organisation of steelmaking capacity and employment by divisions and works in line with the *15 Million Tonnes Proposal*

	Steel making capacity mta		*Number of employees (000)*	
	1979/80, Q4	*1980/81, Q4*	*1979/80, Q4*	*1980/81, Q4*
1 Scottish Division	2.1	2.1	10.6	9.5
Hallside	0.1	—		
Ravenscraig	2.0	2.1		
2 Scunthorpe Division	3.6	3.1	18.0	15.8
3 Sheffield Division	3.8	3.4	32.1	31.4
4 Teeside Division	3.9	3.6	23.3	19.0
5 Welsh Division	5.0	2.7	40.0	28.3
Llanwern	2.3	} 2.7		
Port Talbot	2.3			
Shotton	0.4			
6 Tubes Division	1.4	0.3	26.9	22.9
Clydesdale	0.3	0.3		
Corby BOS	1.0	—		
Corby Arc	0.1	—		
7 Establishments other than iron and steel activities	—	—	15.7	16.4
Total	19.8	15.2	166.6	143.3

Sources: BSC 1979; BSC: Interview

per cent in 1980 to 90 per cent in 1982. Employees were to be further slashed by 20,000 and more rationalisation would be carried out in the following works:

1 Normanby Park was to be closed
2 Appleby–Frodingham No. 1 rod mill was to be closed
3 Lackenby bar mill was to be closed
4 Ebbw Vale 4 stand cold mill was to be closed
5 Distington ingot mould foundry was to be closed
6 Templeborough to be run on a restricted basis
7 Coking capacity to be reduced by closing ovens at Shotton and Hartlepool. Orgreave and/or Brookhouse was to continue until alternative fuel arrangements could be made for works in the Sheffield area
8 Velindre (tinplate) to be reduced to a one-shift operation.

(BSC 1980: 2)

Thus reorganisation and rationalisation in the composition of the Corporation establishments was realised by scrapping many old, small-scale works and factories which were geographically widely spread, and by modernising and enlarging the main works, thereby concentrating production there. The large amount of investment for the *Ten Year Development Strategy* announced in 1973 (based upon the demand forecast which required a liquid steel capacity range of 36–38 million tonnes) would have increased the liquid steel capacity range to 5–6 million tonnes on average for each of the five major integrated iron and steel works, leading to profitability in each process as a result of plant and equipment modernisation, as long as the demand forecast was correct. Unfortunately, the world recession and stagnation and the decline of the UK manufacturing industry forced the Corporation to operate their existing new equipment below full capacity and led them to change their modernisation programme as shown in Table 8.2: the liquid steel capacities of the major works were greatly reduced and these reduced capacities are well under the appropriate capacity required for modern large-scale integrated iron and steel works. In fact, crude steel output has continued to decline since 1972 (24.2 million tonnes) and the rate of this decline has accelerated since 1978 (17.3 million tonnes). The output in 1979 and 1980 decreased to 14.1 million and 11.9 million tonnes respectively, although in 1981 it slightly increased to 14.0 million tonnes. In spite of the considerable plant and equipment modernisation and rationalisation of the Corporation establishments, the Corporation has reduced its market share, allowing private producers and imports to expand (the market share for steel products of the Corporation decreased on average from 68.6 per cent over the period 1967–9 to 48.0 per cent in 1979). This is due to a decline in the Corporation's competitiveness and is revealed in relation to the total share of the UK steel industry and the management of the Corporation. The number of employees decreased markedly from the beginning of the second half of the 1970s and over the period 1976–80 plummeted to 87,000.

THE IMPACT OF MODERNISATION UPON DIVISION OF LABOUR

The general impact of modernisation upon labour was to decrease the number of employees and recreate a division of labour across the BSC works. Employment in those areas where over 80 per cent of the Corporation's employees worked dropped from 175,468 to 134,807 over the period 1972–9.[8] Male employment decreased more than female, losing 24.1 per cent as against a 10.0 per cent decrease for full-time and 11.4 per cent for part-time women workers. The male composition ratio simultaneously fell from 93.0 to 91.9 per cent, while for women, full-timers rose from 5.9 to 6.9 per cent, and part-timers from 1.1 per cent to 1.2 per cent.

The basic framework of division of labour in the Corporation is shown in the four occupational categories given in the manpower statistics: management, technical, clerical, etc. employees; process operatives; maintenance and craftsmen; ancillary employees. The decrease in employment among process operatives is the highest (31.1 per cent), followed by maintenance and craftsmen (24.7 per cent), ancillary employees (22.5 per cent) and management, technical, clerical, etc. employees (12.9 per cent). For males alone, the figures are: process operatives (31.2 per cent), maintenance and craftsmen (24.7 per cent), ancillary employees (24.4 per cent) and management, technical, clerical, etc. employees (13.6 per cent). Female employment suffered decreases in process operatives (15.9 per cent), ancillary employees (11.2 per cent), and management, technical, clerical, etc. employees (9.6 per cent). This is set off to a small extent by a 38.9 per cent increase in maintenance and craftsmen, but the high percentage masks the small numerical rise from seven to twenty-five persons.

Reflecting the above changes in numbers of employees, the percentage total in each category has also changed. In 1972 the highest percentage of the four categories was process operatives (30.9 per cent), followed by maintenance and craftsmen (27.0 per cent), management, technical, clerical, etc. employees (27.0 per cent) and ancillary employees (15.1 per cent); while in 1979 the order changed to management, technical, clerical, etc. employees (30.6 per cent), process operatives (27.7 per cent), maintenance and craftspeople (26.5 per cent) and ancillary employees (15.1 per cent).[9] The changes in percentage of male employees by category are as follows: process operatives (from 32.9 per cent to 29.9 per cent); maintenance and craftsmen (from 29.0 per cent to 28.8 per cent); management, technical and clerical, etc. employees (from 24.0 per cent to 27.4 per cent); ancillary employees (from 14.0 per cent to 13.9 per cent). As for female employees they are as follows: management, technical and clerical, etc. employees (from 65.8 per cent to 66.3 per cent); ancillary employees (from 30.7 per cent to 30.4 per cent); process operatives (from 3.3 per cent to 3.1 per cent); maintenance and craftspeople (from 0.1 per cent to 0.2 per cent).

Management, technical, clerical, etc. employees

The number of employees in this category decreased by 6,118 during the period 1972–9. This total can be broken down into: directors and managers, management in special functions, scientists, technologists; supervisors (inclusive of functional specialists, foremen, clerical supervisors), technicians, clerical, etc. employees (male); supervisors (inclusive of supervisors and technicians), clerical, etc. employees (female). The decrease rates in the above three groups and occupations vary in degree. Supervisors, technicians, clerical, etc. employees (male) decreased by 19.7 per cent and females by 9.6 per cent, while as a whole directors and managers, management in special functions, scientists, and technologists increased by 10.4 per cent.

The percentage of supervisors, technicians, clerical, etc. employees (male) fell from 66.3 to 61.1 per cent, females rose from 17.0 to 17.7 per cent and directors and managers, management in special functions, scientists, technologists also rose from 16.7 to 21.2 per cent. In contrast to a relatively large fall in the percentage share of supervisors, technicians, clerical, etc. employees, that of directors and managers, management in special functions, scientists and technologists rose to over 20 per cent.

Directors, managers, management in special functions, scientists and technologists

The changes in numbers and composition of this category are shown in Table 8.3. In spite of the large reduction in employment in the Corporation as a whole, the number of directors, managers and those

Table 8.3 Number and composition of directors, managers, management in special functions, scientists and technologists

	1972 October–December (person)	(%)	1979 October–December (person)	(%)	% change 1979/1972
Directors	24	0.3	33	0.4	37.5
Managers	114	1.4	689	7.9	504.4
Management in special functions					
Production	1,495	18.9	1,358	15.5	–9.2
Engineering	1,990	25.1	2,173	24.8	9.2
Finance	497	6.3	673	7.7	35.4
Commercial	368	4.6	662	7.6	79.7
Personnel	409	5.2	503	5.7	23.0
Administrative	217	2.7	198	2.3	–8.8
Other	1,087	13.7	1,062	12.1	–2.3
Sub-total	6,063	76.5	6,629	75.7	9.3
Scientists, Technologists	1,721	21.7	1,399	16.0	–18.7
Total	7,922	100.0	8,750	100.0	10.4

Source: Hasegawa 1988
Note: Directors includes the board members and managing directors

in management in special functions increased, which suggests that the managerial structure and hierarchy of the Corporation as a whole expanded. The large jump in the number of managers while equipment modernisation and rationalisation of employment were under way encouraged diversification and sub-division of managerial authority and functions, and thus the necessity for managers increased. The larger number of employees in the commercial, finance, personnel and engineering categories of management in special functions suggests the increased managerial function in the respective departments. Increased management in commercial departments reflects dealings with the increased competition in the world-wide recession in a situation of higher productivity resulting from large plant and equipment investment. Increased management in finance reflects dealings with the continued deficits in business.[10] Increased management in personnel reflects dealings with difficult and complex situations of labour and industrial relations management engendered by employment rationalisation.[11] Increased management in engineering reflects the plant and equipment modernisation based upon a 'scrap-and-build' policy, which aimed to concentrate iron and steel production in the five major works whose equipment was large, continuous, fast and highly automated. The decrease in the number and percentage of employees for management in production and administration reflects the concentration of administrative work generated by plant and equipment modernisation and rationalisation, as well as rationalisation of office work by means of office automation. A sharp decrease in the number and percentage of scientists and technologists suggests less regard for research and development, and increased dependence upon the very modern iron and steel technologies of other advanced nations.[12] In spite of such explanations for changes we cannot dismiss the fact that the increasing trend of directors and managers is due to an apparently irresistible inertia of developing layers of managers, which in consequence creates inefficiency in managerial function.

Supervisors, technicians, clerical employees, etc. (males)

Changes in the number and composition of employees in this category are shown in Table 8.4 and indicate a rise in the total percentage shares of functional specialists, foremen and clerical supervisors from 41.5 per cent to 47.7 per cent, accounting for almost half of the total employees in this category. The rise in percentage share of clerical supervisors in contrast to the large decrease in clerical employees reflects the difficulty of decreasing supervisory staff, thus creating a smaller span of control. The decrease in the number of functional specialists and technicians–mechanical, etc. is relatively

Table 8.4 Number and composition of supervisors, technicians and clerical employees (males)

	1972 October–December		1979 October–December		% change 1979/1972
Functional specialists	4,412	(14.1)	3,791	(15.1)	−14.1
Foreman	7,270	(23.2)	6,203	(24.6)	−14.7
Clerical supervisors	1,332	(4.2)	1,202	(4.8)	−9.8
Technicians:					
Chemical etc.	3,084	(9.8)	2,329	(9.2)	−24.5
Mechanical etc.	954	(3.0)	861	(3.4)	−9.7
Other	578	(1.8)	893	(3.5)	54.5
Draughtsmen	924	(2.9)	510	(2.0)	−44.8
Clerical employees	9,219	(29.4)	6,107	(24.3)	−33.8
Apprentices and trainees	1,580	(5.0)	1,422	(5.6)	−10.0
Others	1,990	(6.3)	1,864	(7.4)	−6.3
Total	31,343	(100.0)	25,182	(100.0)	−19.7

Source: Hasegawa 1988

small and the number of technicians–other has even increased. This trend reflects the formation and increase in occupations concerned with new technical knowledge which appeared along with the mechanisation and computerisation of office jobs. The number of draughtsmen, clerical employees and technicians–chemical, etc. reduced sharply. This reflects the increased use of automation both in process and business operations which resulted from the equipment modernisation, reorganisation and rationalisation of the Corporation's establishments. A relatively small decrease in the number of apprentices and trainees indicates that the large decrease in employment in this category (male supervisors, technicians, clerical employees, etc.) is compensated for by a new and younger workforce.

Supervisors, technicians, clerical employees, etc. (females)

Changes in the number and composition of employees in this category are shown in Table 8.5. Clerical employees, both full and part time, increased markedly, in strong contrast to male clerical employees. Standardisation and simplification of office labour by the use of office automation seem to have made it possible to reduce the number of male clerical employees and replace them by female employees who receive lower wages. The very small decrease in the number of supervisors, technicians, etc. where there is a large increase of clerical employees suggests that the managerial workload of supervisors, technicians, etc. has increased. Typists and machine/telephone operators, whose rate of decrease is dramatic (though they still account for a high percentage share of the total employees in this category) have evidently been affected by office automation. The increase in the number of apprentices and trainees suggests that areas of female jobs expanded and the decrease of supervisors, technicians, etc. and clerical employees is partly compensated for by a young workforce, the same as in the male category. The mixed use of full- and part-time employees in major occupations indicates one of the characteristics of the use of female employees by the Corporation.

Process operative employees

The decrease of process operatives over the period 1972–9 was 16,836. The decrease rate of male operatives was 31.2 per cent, while that of females was 15.9 per cent, only half that of the males. However, in terms of percentage share male operatives are still predominant, with only a slight decline from 99.2 to 99.1 per cent over the same period.

Table 8.5 Number and composition of supervisors, technicians and clerical employees (females)

		1972 October–December		1979 October–December		% change 1979/1972
Supervisor, technicians etc.		1,026	(12.7)	993	(13.6)	−3.2
Clerical	full-time	2,352	(29.2)	2,907	(39.9)	23.6
employees	part-time	35	(0.4)	48	(0.7)	37.1
Typists, machine/	full-time	3,830	(47.5)	2,466	(33.9)	−35.6
telephone operators, etc.	part-time	109	(1.4)	62	(0.9)	−43.1
Apprentices and trainees		413	(5.1)	480	(6.6)	16.2
Others	full-time	255	(3.2)	294	(4.0)	−18.9
	part-time	37	(0.5)	30	(0.4)	15.3
Total		8,057	(100.0)	7,280	(100.0)	−9.6

Source: Hasegawa 1988

Table 8.6 shows decrease rates elsewhere – other production departments and mines and quarries over 65 per cent; steel melting shops and cold reduction mills over 40 per cent; blast furnace and sinter plants, steel foundries and hot rolling mills over 30 per cent; packing and dispatching departments and coke ovens, etc. over 20 per cent; forge and ancillary production plants and ancillary departments – coating, tube, pipe and fittings plants, ancillary departments – and other – over 3 per cent. The only plant type where employment increased was continuous casting plants, where an increase of 85.2 per cent occurred. In terms of percentage share, hot rolling mills have the largest, followed by steel melting shops, blast furnace and sinter plants, tube, pipe and fittings plants and ancillary departments and other, exceeding 3,000 employees. The percentage share of ancillary departments – other and tube, pipe and fittings plants – increased, while that of blast furnace and sinter plants and steel melting shops decreased, but there is no change for hot rolling mills.

Table 8.6 Number and composition of male process operative employees

	1972 October–December		*1979 October–December*		*% change 1979/1972*
Mines and quarries	968	(1.8)	315	(0.9)	−67.5
Coke ovens, etc.	3,428	(6.4)	2,666	(7.2)	−22.2
Blast furnace and sinter plants	6,593	(12.3)	4,056	(11.0)	−38.5
Steel melting shops	8,459	(15.7)	4,261	(11.5)	−49.6
Continuous casting plants	385	(0.7)	713	(1.9)	85.2
Hot rolling mills	13,802	(25.7)	9,501	(25.7)	−31.2
Cold reduction mills	4,845	(9.0)	2,751	(7.4)	−43.2
Ancillary (coating)	1,945	(3.6)	1,775	(4.8)	−8.7
Depts (other)	3,639	(6.8)	3,528	(9.5)	−3.1
Tube, pipe and fittings plants	3,726	(6.9)	3,585	(9.7)	−3.8
Forge and ancillary proc. plants	479	(0.9)	420	(1.1)	−12.3
Steel foundries	501	(0.9)	318	(0.9)	−36.5
Packing and despatching depts	2,287	(4.3)	1,697	(4.6)	−25.8
Other production departments	1,514	(2.8)	488	(1.3)	−67.8
Junior process operatives	1,169	(2.2)	895	(2.4)	−23.4
Total	53,740	(100.0)	36,969	(100.0)	−31.2

Source: Hasegawa 1988

The number of employees in mines and quarries and other production departments whose decrease rates were over 60 per cent was 315 in 1979. The decrease of employment in mines and quarries suggests an increased dependence upon foreign ores and other materials required by the expansion and modernisation of the coastal integrated iron and steel works.[13] The decrease of employment in other production departments suggests the progressive closure of old and small plants and factories. The rate of decrease in employment of ironmaking, steelmaking and rolling processes other than continuous casting plants exceeds 30 per cent. This reflects the decrease in employment due to the modernisation and expansion of the five main works in rendering their production system large, continuous, fast and highly automated, requiring a relatively small workforce; the closure of old and small iron and steel works was also promoted along with such modernisation. The increase in employees in continuous casting plants implies a larger decrease in employees in the ingot, soaking, and blooming processes, which had been rationalised by the continuous casting process. Though their rates of decrease in employment are relatively small and their percentage shares are not high, the percentage shares of ancillary departments – (coating, other), tube, pipe and fittings plants, forge and ancillary production plants and packing and despatching departments have increased. This suggests that the level of mechanisation and automation in these departments and plants was not as high as that in plants and mills of the main process.

Maintenance and craftsmen

The number of maintenance and craftspeople decreased in this period by 11,714. The decrease among males was 24.7 per cent, while females actually increased by 38.9 per cent, but as observed female employment was very small, numbering only twenty-five in 1979. Therefore the percentage share of male employees is predominantly high with only a very slight decrease from 100 per cent to 99.9 per cent over the same period. Changes in the number and composition of male employees will now be examined.

The Corporation broadly classifies maintenance and craftsmen (male) into skilled, semi-skilled, other workers and apprentices. The strategy of rationalisation in this category of employment was, in order: other workers (37.5 per cent), apprentices (21.9 per cent), skilled (19.7 per cent) and semi-skilled (11.0 per cent). In terms of percentage share, apart from the decline in other workers (from 32.8 per cent to 27.2 per cent), all categories have risen, particularly skilled employees, who went from 47.7 to 50.9 per cent. The category of other workers, whose percentage share decreased, includes low wage employees such as brick labourers, craft labourers and others, but the percentage share itself is still high at 27.2 per cent in 1979, second only to skilled employees.

Table 8.7 shows the employment rationalisation of skilled employees whose percentage share is the highest. Among skilled employees, the number who received a direct impact from equipment modernisation and automation (such as blacksmiths, bricklayers and masons, etc.) decreased greatly, those with relatively small influence (such as boilermakers, instrument craftsmen, roll grinders, etc.) increased their percentage shares slightly, and those whose importance increased (such as fitters and electricians) increased their share more than the others. The rise in percentage share of skilled employees resulted from the rise in percentage shares of fitters and electricians, which indicates that plant and equipment modernisation does not require more skilled employees in general. Rather it requires relatively more maintenance

Table 8.7 Number and composition of male skilled employees (number and %)

	1972 October–December		*1979 October–December*		*% change 1979/1972*
Blacksmiths	325	(1.4)	188	(1.0)	–42.2
Boilermakers	1,068	(7.1)	1,435	(7.9)	–10.8
Bricklayers and masons	1,937	(8.6)	1,201	(6.6)	–38.0
Carpenters and joiners	595	(2.6)	405	(2.2)	–38.0
Electricians	4,023	(17.8)	3,426	(18.9)	–14.8
Fitters	6,698	(29.6)	5,796	(3.9)	–13.5
Instrument craftsmen	441	(1.9)	405	(2.2)	–8.2
Machinists	963	(4.3)	656	(3.6)	–31.9
Template makers	29	(0.1)	21	(0.1)	–27.6
Roll turners	401	(1.8)	330	(1.8)	–17.7
Roll grinders	232	(1.0)	193	(1.1)	–16.8
Other turners	739	(3.3)	646	(3.6)	–12.6
Moulders and coremakers	98	(0.4)	70	(0.4)	–28.6
Pattern makers	99	(0.4)	77	(0.4)	–22.2
Plumbers and pipe fitters	934	(4.1)	876	(4.8)	–0.2
Welders and burners	1,482	(6.6)	1,139	(6.2)	–23.1
Other craftsmen	2,016	(8.9)	1,309	(7.2)	–35.1
Total	22,620	(100.0)	18,173	(100.0)	–19.7

Source: Hasegawa 1988

and employees with new skills in the science and technology of modernised equipment, eliminating work based on skills gained from experience.

Ancillary employees

The number of ancillary employees decreased by 5,993 over this period. Male employees fell 24.4 per cent and female full-time employees 11.1 per cent, less than half the male decrease. Female employees of the Corporation are concentrated in ancillary occupations, the next largest number after clerical employees. Female ancillary employees are composed of full- and part-time employees, as are clerical employees. The rate of decrease is highest in catering employees (part-time) at 44.9 per cent, followed by other employees (full-time) 19.0 per cent, catering employees (full-time) 10.2 per cent, and other employees (part-time) 2.2 per cent. In terms of percentage share catering employees (full-time) account for 44.7 per cent and other employees (part-time) 40.4 per cent, together accounting for 85.1 per cent of the total female ancillary employees.

Changes in employment for male ancillary employees are shown in Table 8.8. Reorganisation and rationalisation of Corporation establishments and plant and equipment modernisation have had various

Table 8.8 Number and composition of male ancillary employees (number and %)

	1972 October–December		*1979 October–December*		*1979/1972 % change*
Fuel, power, water supply	3,044	(13.3)	2,012	(11.7)	−33.9
Warehouse stores, manual	1,236	(5.4)	1,012	(5.9)	−18.1
Rail drivers and shunters	3,631	(15.9)	2,391	(13.9)	−34.1
Road drivers	1,121	(4.9)	1,104	(6.6)	1.7
Mobile plant operators	2,033	(8.9)	1,394	(8.1)	−31.4
Crane drivers and slingers	6,417	(28.1)	5,603	(32.5)	−12.7
Catering employees	139	(0.6)	108	(0.6)	−22.3
General labourers	2,043	(9.0)	1,061	(6.2)	−48.1
Other employees	3,160	(13.8)	2,529	(14.7)	−20.0
Total	22,824	(100.0)	17,250	(100.0)	−24.4

Source: Hasegawa 1988

influences upon ancillary employees as well. The characteristic use of female employees of the Corporation is also seen in this ancillary category. The rate of decrease for male employees is more than double that of female employees, caused by the sharp decline in male employment such as general labourers and rail drivers and shunters, etc., which were affected strongly by the corporate strategy of reorganisation and rationalisation. The rate of decrease in employment for occupations which do not yield much economic benefit from equipment modernisation and automation, such as crane drivers and slingers, and road drivers is small and they have increased their relative importance in terms of percentage share.

LEVELS AND DIFFERENTIALS OF SALARIES AND WAGES

The general condition and its change in the level and differentials in salaries and wages shows us a unique characteristic of management and social relations in the British steel industry. Wages and salaries are an indicator of how employees are organised and integrated, and how they are related to each other in the same organisation.

Employees in the iron and steel activities

The levels of salaries and wages paid to Corporation employees (exclusive of members of the board, directors, managers, management in special functions, scientists and technologists) rose remarkably over the period under study, with a higher increase rate for female employees (220.8 per cent) than males (185.0 per cent). The pay differentials between men and women narrowed; women's pay stood at 49.7 per cent that of men's salaries and wages in 1972, but rose to 57.1 per cent in 1979.

For male employees, wages rose by 190.0 per cent for ancillary employees, 189.1 per cent for supervisors, technicians, clerical employees, etc., 181.7 per cent for process operatives and 177.8 per cent for maintenance and craftsmen. This exacerbated the shortfall in the blue-collar (process operatives, maintenance and craftsmen employees) wage, which stood at 93.7 per cent of that of supervisors, technicians, clerical employees, etc. in 1972, but declined to 91.9 per cent in 1979. In 1972 other comparisons to the wage level of supervisors, technicians, clerical employees, etc. were 96.6 per cent for process operatives, 95.0 per cent for ancillary employees, and 89.6 per cent for maintenance and craftspeople; in 1979 these had become 94.2 per cent, 95.3 per cent, and 86.1 per cent respectively, causing some change in ranking order as well. The gap increased for process operatives and maintenance and craftsmen (whose numbers are large) and decreased for ancillary employees (whose numbers are small).

On the whole, women received greater wage enhancements: 242.9 per cent for process operatives, 218.8 per cent for supervisors, technicians, clerical employees, etc., 217.5 per cent for maintenance and craftspeople and 207.5 per cent for ancillary employees. This resulted in blue-collar wages rising against those of supervisors, technicians and clerical employees, from 83.7 per cent to 84.9 per cent. The wage level of maintenance and craftspeople to that of supervisors, technicians and clerical employees in 1972 was 87.6 per cent, that of process operatives 84.2 per cent and ancillary employees 79.7 per cent, while in 1979 operatives rose to 90.5 per cent, maintenance and craftspeople decreased to 87.3 per cent and ancillary employees decreased to 76.9 per cent. The gap was narrowed for process operatives, whose numbers are very small, but widened for maintenance and craftspeople and ancillary employees, who account for over 30 per cent of employees. Thus it might be concluded that the wage gap between blue-collar workers and supervisors, technicians, clerical employees, etc., in effect widened.

The above trend indicates the following points of managerial strategy:

1 although the wage gap between male and female employees narrowed, the overall level of female wages remains just under 60 per cent of male wages;
2 the average wage levels of both male and female supervisors, technicians and clerical employees remain higher than those of process operatives, maintenance and craftsmen and ancillary employees;
3 the wage gap between male blue-collar workers and male supervisors, technicians, clerical employees, etc. widened and the gap is larger for maintenance and craftsmen, followed by process operatives and ancillary employees;
4 the wage gap between the majority of female blue-collar workers and female supervisors, technicians and clerical employees widened as well, allowing for a decreased rate of wages and a large amount of employment for ancillary employees.

Board members, managers and supervisors[14]

The rates of salary increase of board members, managers and supervisors over the same period were: 184.9 per cent for supervisors, 172.2 per cent for all managers exclusive of the top 0.7 per cent management,

60.3 per cent for the top 0.7 per cent management, and 35.6 per cent for board members; thus the lower the rank in the managerial hierarchy, the higher the salary increase.[15]

Such differences have influenced the levels of salaries as well. The level of salary for the top 0.7 per cent of management as against board members rose from 63.9 per cent to 75.5 per cent over the same period, while that of all managers (other than the 0.7 per cent) rose from 14.5 per cent to 29.1 per cent and that of supervisors rose from 10.7 per cent to 22.4 per cent. The trend of salaries for board members, managers and supervisors, whose percentage share of Corporation employment increased, indicates that the level of salaries for the majority of managers, whose number increased remarkably, is a little less than 30 per cent of that of board members, and the gap against that of the top 0.7 per cent of management is also large. The gap between the level of salaries for supervisors and all managers excluding the top 0.7 per cent of management is not so large, and even narrowed over the same period.

Supervisors, technicians, clerical employees, etc. (males)

Trends in salaries for supervisors, technicians, clerical employees, etc. (males) are shown in Table 8.9. The trend of salaries shown in the table indicates that the level of salary for functional specialists as against

Table 8.9 Salaries paid to supervisors, technicians, clerical employees (males, £/week)

	1972 October–December	*1979 October–December*	*% change 1979/2*
Functional specialists	40.1 (106.1)	112.5 (108.0)	180.5
Foremen	45.4 (120.1)	134.5 (129.1)	196.3
Clerical supervisors	37.8 (100.0)	104.2 (100.0)	175.7
Technicians:			
Chemical etc.	35.4 (93.7)	110.5 (106.0)	212.1
Mechanical etc.	41.4 (109.5)	122.9 (117.9)	196.9
Other	35.9 (95.0)	101.9 (97.8)	183.8
Draughtsmen	37.0 (97.9)	101.2 (97.1)	173.5
Clerical employees	32.2 (85.2)	92.8 (89.1)	188.2
Apprentices and trainees	18.9 (50.0)	47.5 (45.6)	151.3
Others	32.9 (87.0)	104.4 (100.2)	217.3

Source: Hasegawa 1988
Note: ( ) Indicates the percentage of salaries paid for various occupations as compared to clerical supervisors

foremen declined from 88.3 to 83.6 per cent, while clerical supervisors declined from 83.3 to 77.5 per cent, thus raising the relative level of salary for foremen compared with other supervisory employees. The level of salaries for clerical supervisors and functional specialists showed a relative decline, but still remained at the medium ranking. The salaries of technicians, chemical and mechanical, showed a relative increase exceeding those of functional specialists and clerical supervisors. Although the salary of clerical employees rose relatively, the level is the next lowest to that of apprentices and trainees. The salaries of draughtsmen and apprentices and trainees declined most severely and in particular that of the latter accounts for less than half that of clerical supervisors in 1979.

Supervisors, technicians, clerical employees, etc. (females)

Trends in salaries for female supervisors, technicians, and clerical employees, etc. are shown in Table 8.10. The trends shown in the table indicate that the level of salaries for all occupations, including low-ranking part-time employees, moved closer to that of supervisors, technicians, etc., thus raising general salary level. The small increase rate of salaries for full-time typists, machine/telephone operators, etc. and full-time clerical employees (in which the majority of female employees are concentrated) suggests that the relative position of these female employees has declined. The relative increase of part-timers' wages

Table 8.10 Salaries paid to supervisors, technicians, clerical employees (females, £/week)

		1972 October–December	*1979 October–December*	*% change 1979/1972*
Supervisors, technicians, etc.		27.8 (100.0)	84.6 (100.0)	204.3
Clerical employees	full-time	20.5 (73.7)	64.0 (75.7)	212.2
	part-time	11.7 (42.1)	43.9 (51.9)	275.2
Typists, machine/ telephone ops, etc.	full-time	20.6 (74.1)	63.6 (75.2)	208.7
	part-time	12.4 (44.6)	46.1 (54.5)	271.8
Apprentices and trainees		12.3 (44.2)	42.5 (50.2)	245.5
Other	full time	19.8 (71.2)	67.1 (79.3)	238.9
	part time	10.5 (37.8)	49.4 (58.4)	370.5

Source: Hasegawa 1988
Note: ( ) indicates the percentage of salaries paid for various occupations as compared to supervisors, technicians, etc

reflects the management inclination to make more use of part-timers. Although the level of salary for supervisors, technicians, etc. declined relatively, there is no other occupation with a higher salary level.

Process operatives

Wages went up by 181.7 per cent for male process operatives and 242.9 per cent for female operatives, so that women's wages rose from 49.3 to 60.0 per cent of the male level. As the number of female operatives is small (only 343 in 1979), the wages of male operatives, by plant, will be examined in Table 8.11. Such trends in wages as shown in Table 8.11 indicate that the higher the level of automation the higher the relative decline in wages. The wage gap between process operatives and

Table 8.11 Wages paid to male process operative employees (£/week)

	1972 October–December	*1979 October–December*	*% change 1979/1972*
Foreman	45.4 (100.0)	134.5 (100.0)	196.3
Mines and quarries	28.8 (63.4)	106.9 (79.5)	271.1
Coke ovens etc.	32.9 (72.5)	97.0 (72.1)	194.8
Blast furnace and sinter plants	33.4 (73.6)	100.0 (74.3)	199.4
Steel melting shops	37.8 (83.3)	104.5 (77.7)	176.5
Continuous casting plants	36.1 (79.5)	114.0 (84.8)	215.8
Hot rolling mills	34.4 (75.8)	88.0 (65.4)	155.8
Cold reduction mills	35.3 (77.8)	95.1 (70.7)	169.4
Ancillary depts:			
Coating	36.6 (80.6)	98.6 (73.3)	169.3
Other	35.0 (77.1)	105.3 (78.3)	200.9
Tube, pipe and fittings plants	32.9 (72.5)	104.9 (78.0)	218.8
Forge and ancillary proc. plants	38.4 (84.6)	122.0 (90.7)	217.7
Steel foundries	34.2 (75.3)	92.3 (68.6)	169.9
Packing and despatching depts	37.9 (83.5)	108.0 (80.3)	185.0
Other production departments	36.0 (79.3)	105.0 (78.1)	191.7
Junior process operatives	11.9 (26.2)	35.6 (26.5)	199.2
Average, exclusive of foreman	34.5 (76.0)	97.2 (72.3)	181.7

Source: Hasegawa 1988
Note: ( ) indicates the percentage of wages paid for various occupations as compared to foreman

foremen became larger mainly because of the small rate of increase in wages for operatives in the key production processes where the level of equipment modernisation and automation is high. This in effect indicates the rise of the relative importance of foremen in the face of changes in labour and management under the progress of automation. We can acknowledge that the wages of operatives for jobs in processes newly created by plant and equipment modernisation are set at the highest level.

Maintenance workers and craftsmen

The rate of wage increase for male maintenance and craftsmen (skilled, semi-skilled, other workers, apprentices) was 180.1 per cent, while that of female employees was 217.5 per cent. Although the increase of female employees was higher than that of male employees, the number of female maintenance and craftspeople was very small, being eighteen in 1972 and twenty-five in 1979. Therefore the lessening of the wage gap between male and female employees is not as significant as that of process operatives. The level of wages for female employees as compared to skilled workers rose from 42.3 per cent to 49.0 per cent over the same period, but is still low.

Among male maintenance and craftsmen, the rate of wage increase for skilled workers was 174.6 per cent, semi-skilled workers 171.4 per cent, other workers 180.1 per cent and apprentices 194.4 per cent. The rate of increase for low-waged workers such as apprentices and other workers is higher than that of skilled and semi-skilled workers, and the rate of increase for skilled workers is higher than that of semi-skilled workers. The wage gap of other workers, apprentices and female workers as compared to skilled workers was narrowed, while that of semi-skilled workers became wider. Table 8.12 shows the wages of skilled workers, which account for about 51.0 per cent of all the maintenance workers and craftspeople. The trends indicate the following points: parity among maintenance and craftspeople improved, in that the wage gap between male and female employees became smaller and the relative wages of low-waged employees (such as 'other workers' and apprentices) were increased, while on the other hand the relative wages for semi-skilled workers declined. The wages for occupations (such as electricians and fitters) which require new technical knowledge in addition to the existing skills gained from experience, rose substantially and remained at this high level. Occupations where work became simplified and standardised due to plant and equipment modernisation and automation (for example, instrument craftspeople) had to suffer a sharp decline in wages. The wages of highly skilled occupations based upon experience (such as bricklayers and masons, template makers, roll turners and roll grinders) remained high. The relative rise in the wage of foremen as compared to other skilled workers suggests an increase in their responsibilities and managerial ability, thus making it possible to increase their managerial function towards workers.

Ancillary employees

The rate of wage increase for male ancillary employees was 190.0 per cent, while that of female full-time was 207.5 per cent and female part-time 220.0 per cent. The wage level of female employees compared to that of male employees over the same period rose from 47.5 per cent to 50.4 per cent for full-time employees and from 28.0 per cent to 30.9 per cent for part-time employees. We can acknowledge that the wage

Table 8.12 Wages paid to male skilled employees (£/week)

	1972 October–December	*1979 October–December*	*% change 1979/1972*
Foremen	45.4 (100.0)	134.5 (100.0)	196.3
Blacksmiths	38.7 (85.2)	107.2 (79.7)	177.0
Roller makers	43.5 (95.8)	112.0 (83.3)	157.5
Bricklayers and masons	45.0 (99.1)	123.3 (91.7)	174.0
Carpenters and joiners	39.2 (86.3)	108.8 (80.9)	177.6
Electricians	41.9 (92.3)	118.0 (87.7)	181.6
Fitters	42.1 (92.7)	116.7 (86.8)	177.2
Instrument craftsmen	41.1 (90.5)	108.6 (80.7)	164.2
Machinists	38.2 (84.1)	110.8 (82.4)	190.1
Template makers	49.9 (109.9)	123.0 (91.4)	146.5
Roll turners	42.5 (93.6)	116.2 (86.4)	173.4
Roll grinders	44.9 (98.9)	116.5 (86.7)	159.5
Other turners	37.6 (82.8)	101.6 (75.5)	170.2
Moulders and core makers	28.9 (63.7)	99.0 (73.6)	242.6
Pattern makers	36.3 (80.0)	112.1 (83.3)	208.8
Plumbers and pipe fitters	41.3 (91.0)	113.8 (84.6)	175.5
Welders and burners	41.5 (91.4)	115.4 (85.8)	178.1
Other craftsmen	41.1 (90.5)	106.4 (79.1)	158.9
Average, exclusive of foreman	41.8 (92.1)	114.8 (85.4)	174.6

Sources: Hasegawa 1988
Note: () indicates the percentage of wages paid for various occupations as compared to foreman

gap between the sexes became narrower and the rate of increase higher for those whose wage is lower. Considering the rate of inflation the relative increase of female employees' wages was inevitable, but while the difference in remuneration between male and female narrowed, the female wage level remained extremely low.

The wages of male ancillary employees are shown in Table 8.13. The wage trends indicate that in the category of ancillary employees not immediately influenced by plant and equipment modernisation and automation the wage gap between male and female employees narrowed and a relatively high wage increase among the lowest wage occupations occurred. Occupations whose importance increased due to plant and equipment modernisation and automation (such as fuel, power and water supply employees and mobile plant operators) have a relatively higher increase rate of wages. Occupations such as rail drivers and shunters, road drivers and crane drivers, whose skills gained from physical experience were not affected by plant and equipment modernisation and automation, are guaranteed a high remuneration even though their relative position declined. The wages for warehouse stores, manual and catering employees not directly influenced by plant and equipment modernisation and automation have relatively decreased. The relative increase in the wage of foremen was observed in this category of ancillary employees as well, suggesting the rise in authority and importance of foremen vis-à-vis general workers.

Table 8.13 Wages paid to male ancillary employees (£/week)

	1972 October–December	*1979 October–December*	*% change 1979/1972*
Foreman	45.4 (100.0)	134.5 (100.0)	196.3
Fuel power, water supply	34.2 (75.3)	101.3 (75.3)	196.2
Warehouse stores manual	32.2 (70.9)	92.4 (68.7)	187.0
Rail drivers and shunters	35.5 (78.2)	102.1 (75.9)	187.6
Road drivers	35.8 (78.9)	102.2 (76.0)	185.5
Mobile plant operators	36.9 (81.3)	107.1 (79.6)	190.2
Crane drivers and slingers	39.6 (87.2)	108.1 (80.4)	173.0
Catering employees	25.4 (55.9)	68.8 (51.2)	170.9
General labourers	22.6 (49.8)	74.3 (55.2)	228.8
Other employees	26.3 (57.9)	78.0 (58.0)	196.6
Average, exclusive of foreman	33.9 (74.7)	98.3 (73.1)	190.0

Source: Hasegawa 1988
Note: ( ) indicates the percentage of wages paid for various occupations as compared to foreman

CONCLUSION

The modernisation of the British steel industry went hand-in-hand with a general rationalisation. As such, although the ultimate goal in both British and Japanese modernisation was the realisation of state-of-the-art production systems, certain distinctive features emerged in the British experience. One of these was the release of a large part of the existing workforce through the closure of older plants and steelworks. At the same time, division of labour was restructured and the remuneration system reworked; this can be interpreted as managerial strategy, itself influenced by contemporary technology, influencing social relations. This, in Britain, is the circumstance of management and labour at the stage of contrived compatibility.

9 Management and labour in British steel works

Two factors, modernisation and rationalisation, lay behind the changes that occurred at works level in the British steel industry, and the Scunthorpe steel works was chosen for an examination of their effect on management and labour.[1]

Just as in Japan, the goal of modernisation in Scunthorpe was the establishment of a large-scale, continuous, high-speed production system with the extensive use of automation. Yet the employment and work practices that emerge in this detailed study of the works indicate a quite different situation from that of Japan. This chapter presents data that will later be used to show how the changes relate to the main theme of our study, i.e. relative advance and convergence.

MODERNISATION OF SCUNTHORPE WORKS

The Scunthorpe works is a representative large-scale integrated works of the British Steel Corporation. This was formerly the Appleby–Frodingham works belonging to the United Steel Company before the 1967 nationalisation. The plant and equipment modernisation of this works was completed in 1973 and was the successor to the pre-nationalisation programme known as the Anchor Project.[2] The total works area is 1,700 acres (some 688 hectares), with an ironmaking capacity of 2,900,000 tonnes per year, crude steelmaking capacity of 4,900,000 tonnes per year and actual crude steel output of 3,100,000 tonnes (1983). The main products are plates, sections, rods and slabs. The number of employees in 1981 was 10,203 and the rate of sub-contracting 10 per cent (1983). Although this is a coastal works, the works site is situated 35.2 kilometres (22 miles) away from the Immingham Bulk Terminal.

Ironmaking process

The main equipment used for this process is a blast furnace. There used to be ten blast furnaces at the Scunthorpe works, but most of them were old and small. There are now four blast furnaces in regular use, two built in 1939 with a hearth diameter of 8.3 metres and two built in 1954 with a hearth diameter of 9.4 metres. Each furnace has a three-hour cast cycle and one tap hole. The resultant iron is transferred to the steel plant by a torpedo car with a 250 tonne capacity. There are no newly-built blast furnaces in use, but the charging of raw materials is thoroughly automated.

Steelmaking process

The main equipment modernisation of this process was the replacement of open hearth furnaces by LD converters and the introduction of continuous casters. In 1973 three 300-tonne basic oxygen converters were installed and the steelmaking capacity was 4,500,000 tonnes per year. The average charge-to-tap time is thirty-five minutes and the control in the vessel is automated. The continuous caster is a machine used to streamline the former separate ingot-making, soaking and blooming processes into one process. The two slab casting machines equipped in 1974 can produce four sizes of slab between 230 mm thick and 1,270 mm and 1,830 mm wide.

Rolling process

The rolling process includes the plate mill, the rod mill, the medium section mill and the heavy section mill. The plate mill was equipped in 1958 and it is on-line computer-controlled from order acceptance to final dispatch. The rod mill was installed as an extension of the Anchor Project, and it has an annual capacity of 600,000 tonnes and is also computer-controlled. The medium section mill was installed in 1973 and with its computer control systems is one of the most advanced in Europe. Its annual capacity is 475,000 tonnes. The heavy section mill is an old machine installed in 1949 and is used to produce piling, universal beams, channels and columns.

Automation

Eight process computers are installed, one in each of the following processes: blending and stocking, blast furnace charging and operations, steelmaking, continuous casting, bloom and billet mill, medium section mill, plate mill and rod mill. Five central mainframe computers are installed: two dealing with accounting and finance, and three for production control and order acceptance through to shipping. In addition they are streamed into head office computers covering finance, accounting and salaries, ore control, stores system and personnel data. In the offices micro-computers and word processors are being introduced.

Although the Scunthorpe works is a coastal, modern, integrated works, it was constructed through the modernisation and enlargement of an old works; the result of this is that the substance of modernisation is inferior in the total site layout, distance from the coast and the level and range of equipment modernisation in and between processes in comparison to the modern works of the Japanese large-scale iron and steel enterprises. Despite the computerisation of the main processes and administrative areas, its level is not as high as in the Japanese works. For example, the Hirohata works of the Nippon Steel Corporation, whose crude steel capacity is almost the same as Scunthorpe, has twenty-three process and seven business computers, while the modern Kimitsu works of the same corporation has four process computers just in one plate mill. Since 1980 sub-contracting has been pursued as a part of employment rationalisation but not to a great extent.[3] Whether the in-works sub-contracting system as found in Japan will develop at Scunthorpe seems to depend upon the implementation of future plant and equipment modernisation, changes in industrial relations and the growth of sub-contracting firms.

CHANGES IN WORKFORCE COMPOSITION AND MANAGEMENT

The modernisation of the works is centred upon the enlargement of various items of capital equipment and their continuous, high-speed and automated operation in the three main processes; ironmaking, steelmaking and rolling. In consequence, its impact is greatest upon operators in these processes, but it will also inevitably extend to workers in other processes.

Table 9.1 shows changes in the number and composition of the Scunthorpe workforce by occupational group. The total number of employees slightly increased over the period 1972–9, but sharply decreased in 1981. This can be ascribed to the choice of Scunthorpe as a core works of the British Steel Corporation in the modernisation

Table 9.1 Number and composition of employees at Scunthorpe works (numbers and per cent)

	December 1972		*December* 1979		*March* 1981	
Managerial, technical, clerical employees	4,566	(24.3)	5,076	(26.9)	3,426	(28.5)
Process operatives	6,471	(34.5)	5,833	(30.9)	3,132	(26.1)
Maintenance and craftsmen	4,955	(26.4)	5,527	(29.9)	4,124	(34.3)
Ancillary employees	2,776	(14.8)	2,469	(13.1)	1,324	(11.0)
Total	18,768	(100.0)	18,905	(100.0)	12,006	(100.0)

Source: Scunthorpe works (BSC): Interview

Note: Maintenance and craftsmen include skilled, semi-skilled, other workers and apprentices. Ancillary employees include employees at fuel, power, water supply, warehouse stores manual, rail drivers and shunters, road drivers, mobile plant operators, crane drivers and slingers, catering employees, general labourers and other employees

programme of the late 1970s, followed by the effects, broadly similar to that experienced elsewhere, of subsequent modernisation and rationalisation. The occupational groups in which employment continuously decreased over this period are process operatives and ancillary employees; the employment of management, technical and clerical employees and maintenance and craftsmen employees increased slightly at first, but began to decline later. In terms of the composition rate, that of management, technical and clerical employees and maintenance and craftsmen employees continued to increase throughout this period, while that of process operatives and ancillary employees dropped. Consequently, process operatives, who were placed first in composition in 1972, dropped to third place in 1981, while maintenance and craftsmen moved from second to first place and management, technical and clerical from third to second.

The trend in employment composition detailed in Table 9.1 shows its characteristics all the more clearly when supplemented by data preceding this period. L.C. Hunter in 1970 presented information on employment composition in the industry over the period from 1957 to 1966; the composition percentage of managerial, technical and clerical employees increased from 14.9 to 21.0 per cent, and general and maintenance employees also increased from 31.5 to 32.7 per cent, while that of process workers decreased from 53.6 to 46.2 per cent (Hunter *et al.* 1970: 125). The continuity in trend recognised above is found in the case of the Scunthorpe works and it should be noted that the percentage of managerial, technical and clerical employees has gone up as remarkably as that of process operatives has gone down.

Factors contributing to the changes in the number and composition of employees in the various occupational groups can be examined using data covering all of the British Steel Corporation. In the case of process operatives, the decrease in their number and composition reflects the sizeable decrease in employment in mines and quarries, steel melting shops, blast furnace and sinter plants and rolling mills consequent upon equipment modernisation and automation in the ironmaking, steelmaking and rolling processes. In the case of the ancillary employees group, the decrease in their number and composition reflects plant and equipment modernisation and automation in the areas of fuel, power and water supply, employment of rail drivers and shunters, and mobile plant operatives. In addition, since 1980 the decrease in the number of general labourer, catering and other employee positions, which were mainly filled by sub-contracted female employees, is reflected in the changes in the number and composition of the ancillary employees group. For maintenance and craftsmen (skilled, semi-skilled, other workers, apprentices), the number of employees has decreased, but the composition percentage has increased; in particular the rise and high level of skilled employees (from 47.7 per cent to 50.9 per cent) should be noted. Among skilled employees, the rise in percentage and the high level of electricians and fitters (only two occupations account for 50.8 per cent of all skilled employees) should be noted and this reflects the rise in the relative importance of this occupational group as a whole. Lastly, although the number of managerial, technical and clerical employees has decreased, the composition ratio has risen. In particular, the higher percentage of directors, managers and scientists and technologists, as well as that of supervisors, technicians, clerical employees (female) is of note, and this is reflected in the changes in the composition rate of this occupational group as a whole.

Examination of the relationship between the occupational groups and organisation of the works shows that the majority of maintenance and craftsmen, whose composition rate was the highest in 1981, are engaged in manual work of a technical nature, the second highest group – managerial, technical and clerical employees – are engaged in these functions in all departments (see Table 9.4), process operatives in manual work of line organisation, and ancillary workers in manual work of all kinds.

Age and length of service

Table 9.2 shows the composition by age in 1981, which we can use to examine the influence of plant and equipment modernisation and

Table 9.2 Composition of employees by age at Scunthorpe works, 1981 (numbers and per cent)

	<20	*20–29*	*30–39*	*40–49*	*50–59*	*over 60*	*Total*
Managerial, technical, clerical employees	115	645	901	789	562	15	3,027
	(3.7)	(21.3)	(29.8)	(26.1)	(18.6)	(0.5)	(100.0)
Process operatives	233	817	639	645	526	35	2,895
	(8.0)	(28.2)	(22.1)	(22.3)	(18.2)	(1.2)	(100.0)
Maintenance and craftsmen	228	1,021	641	594	641	35	3,160
	(7.2)	(32.3)	(20.3)	(18.8)	(20.3)	(1.1)	(100.0)
Ancillary employees	189	208	254	235	228	10	1,121
	(16.9)	(18.6)	(22.7)	(21.0)	(20.3)	(0.9)	(100.0)
Total	765	2,691	2,435	2,263	1,957	95	10,203
	(7.5)	(26.4)	(23.9)	(22.2)	(19.2)	(0.9)	(100.0)

Source: Scunthorpe works (BSC): Interview

Table 9.3 Composition of employees by length of service at Scunthorpe works, 1981 (numbers and per cent)

	<4	*5–9*	*10–19*	*20–29*	*30–39*	*over 40*	*Total*
Managerial, technical, clerical employees	441	547	956	735	313	35	3,027
	(14.6)	(18.1)	(31.6)	(24.3)	(10.3)	(1.2)	(100.0)
Process operatives	1,083	546	583	512	138	33	2,895
	(37.4)	(18.9)	(20.1)	(17.7)	(4.8)	(1.1)	(100.0)
Maintenance and craftsmen	1,156	780	668	359	176	21	3,160
	(36.6)	(24.7)	(21.1)	(11.4)	(5.6)	(0.7)	(100.0)
Ancillary employees	475	339	196	76	30	5	1,121
	(42.4)	(30.2)	(17.5)	(6.8)	(2.7)	(0.4)	(100.0)
Total	3,155	2,212	2,403	1,682	657	94	10,203
	(30.9)	(21.7)	(23.6)	(16.5)	(6.4)	(0.9)	(100.0)

Source: Scunthorpe works (BSC): Interview

rationalisation. The average age composition of the whole workforce breaks down to those under 29 years old accounting for 33.9 per cent, while those over 40 account for 42.3 per cent. It can thus be said that the age composition as a whole is rather high. When the age composition of the white-collar employees – managerial, technical and clerical – is compared with that of the other three blue-collar occupational groups, we find that in the former those under 29 years old accounted for 25.0 per cent and those over 40 for 45.2 per cent, while in the latter those under 29 accounted for 37.6 per cent and those over 40 for 41.1 per cent, indicating the higher age composition among white-collar workers. Among the blue-collar occupational groups the highest percentage of the under-29 age group appears among maintenance and craftsmen, followed by process operatives and ancillary employees; the over-40 age group figures most, in decreasing order, among ancillary employees, process operatives and maintenance and craftsmen.

In the examination of the average length of service of the whole workforce (see Table 9.3), those of less than nine years' service accounted for 52.6 per cent, while those with more than twenty years' service account for 23.8 per cent; a rather shorter length of service is thus predominant. Comparing white- and blue-collar employees, the former accounted for 32.7 per cent having less than nine years' service and 35.8 per cent with over twenty years' service, while the latter groups registered 61.0 and 18.8 per cent respectively. Length of service was hence much shorter among blue-collar employees, and among their occupational groups ancillary staff, maintenance and craftsmen and process operatives ranked in order in terms of employees with less than nine years' service, while the corresponding placing for those with over twenty years' service was process operatives, maintenance and craftsmen and ancillary employees.

The analysis in Table 9.3 indicates that while the age composition figures are rather high as a whole, those for length of service are rather low. This suggests the existence of a large number of employees recruited in the middle of their working life.[4] The levels of age and length of service in terms of composition are higher for white-collar employees than blue-collar, suggesting that any trend of long-term employment is found more among white-collar employees. Among blue-collar workers, age composition is almost the same for every occupational group, but length of service is relatively high among process operatives, for whom the seniority system was established, and also for maintenance and craftsmen, who are organised in craft unions[5] and composed of apprentices and semi-skilled and skilled employees whose wage level is occupationally fixed.

Managerial employees by direct and indirect departments

Equipment modernisation and rationalisation requires an enhanced managerial function. Table 9.4 shows a comparison of managerial composition by direct and indirect departments as analysed between 1978 and 1981. The largest change that occurred in this period is that the managerial employees of direct departments declined drastically, while those of indirect departments increased markedly. Among the latter, an organisation in which the composition level rose significantly was the engineering department; this shows that technical functions, until then the preserve of the direct departments, were transferred to the engineering department, and the division of function between the direct and indirect departments and within the establishment developed along with plant and equipment modernisation.

The order in percentage composition for 1974 was personnel in the highest position, followed by technical and commercial departments; and in 1981, the order was engineering, finance, metallurgy, personnel and social policy departments. The high percentage of the engineering department in 1981 is surprising and notable. Metallurgy had not existed as an independent section in 1974, but due to the enlargement and strengthening of the engineering department it seems to have increased its importance. The finance department existed as an accounting/administration department in 1974, but in 1981 was changed to the finance department, thus increasing its function when the Scunthorpe works became the central unit of the General Steels Group under the new management organisation based upon product-based businesses.[6] The personnel and social policy department existed as a personnel department in 1974, but in 1981 a social policy function was added to it, as the importance of industrial relations at establishment level had increased following drastic rationalisation at the Scunthorpe works.[7] Also in 1981 the function of the supply/production control department was increased. The commercial/general sales department is the only department which lowered its percentage in composition. This suggests that this function was partly transferred to the head office of the business group to which the Scunthorpe works belongs.

We can find remarkable changes over the same period in the hierarchy of managerial employees. The managerial hierarchy at Scunthorpe works can be classified into five groups:

1 top management (board members, managing directors)
2 senior management (directors, blue card)
3 management (red card)
4 middle management
5 supervisory positions.

Table 9.4 Composition of managerial employees by direct and indirect departments (Scunthorpe works) (number and per cent)

Managerial hierarchy	*1974 Senior management*				*Managerial hierarchy*	*1981 Senior management*			
Organisation	*Director*	*Blue card*	*Red card*	*Total*	*Organisation*	*Director*	*Blue card*	*Red card*	*Total*
In-direct department					In-direct department				
Accounting and administration	1	4	—	5 (2.8)	Finance controller	1	3	11	15 (10.3)
Personnel	1	8	4	13 (7.3)	Personnel and social policy	1	4	8	13 (8.9)
Technical	1	6	—	7 (3.9)	Engineer	1	8	34	43 (29.5)
Operations and supplies	1	3	—	4 (2.2)	Supplies and production control	1	1	10	12 (8.2)
Commercial	1	6	—	7 (3.9)	Commercial and sales	1	3	—	4 (2.7)
Sub-total	5	27	4	36 (20.2)	Metallurgist	1	3	11	15 (10.3)
Direct department									
Lancashire works	1	6	—	7 (3.9)	Sub-total	6	22	74	102 (69.9)
					Direct department				
Normanby Park works	1	6	32	39 (21.9)	Lancashire/Shelton works	—	1	3	4 (2.7)
Appleby–Frodingham works	1	9	86	96 (53.9)	Scunthorpe works	1	6	33	40 (27.4)
Sub-total	3	21	118	142 (79.8)	Sub-total	1	7	36	44 (30.1)
Total	8 (4.5)	48 (27.0)	122 (68.5)	178 (100.0)	Total	7 (4.8)	29 (19.9)	110 (75.3)	146 (100.0)

Source: Scunthorpe works (BSC): Interview

Note: (a) The Scunthorpe works was a product-based profit centre organisation in 1974, but due to reorganisation was in 1981 an establishment of one operating group in charge of general steels

(b) Managerial hierarchy in 1974 was classified using the classification of hierarchy in 1981 (director, blue card, red card)

(c) ( ) is composition percentage

The following mostly relates to senior management and management employees. First, the number of managerial employees decreased by thirty-two over the same period while the total decrease of employment at the works was about 6,800. This implies a lessening of the control span per managerial staff member from 105 to eighty-two, thus creating the conditions for tighter personnel management. As to changes in managerial positions, the percentage of directors increased slightly, that of blue card managers decreased greatly and that of red card managers increased. The reason for the insignificant change in the percentage of directors is that there was not much change in the number of basic functions of the works. The decrease in the percentage of blue card managers suggests that at this level of the hierarchy the horizontal spread of management function was rationalised, resulting in the concentration of responsibility and authority in fewer blue card managers.

The relative increase of red card managers during the period suggests that at this level the managerial structure has expanded relatively both horizontally and vertically. This trend can be observed in all organisations of the indirect department, and in particular in the engineering department, suggesting the expansion of the managerial structure in this department. On the other hand, in every factory of the direct department, managerial personnel at this level of the hierarchy decreased in number, resulting in the concentration of responsibility and authority. The emphasis of management shifted from direct to indirect functions, which can be considered a result of plant and equipment modernisation.

When compared with the managerial organisation of a Japanese modern steel works like Kimitsu, it differs in that the functions of sales and finance are not absorbed into head office, and works organisation is not centred on direct departments. That is to say, the Scunthorpe works differs from Kimitsu in that the system development and equipment departments do not exist separately and independently of the engineering department. Also in the direct departments, ironmaking and steelmaking and rolling (by products) organisations are not formed as independent organisations. As regards control span, the decrease in the number of employees per manager changed the situation towards a reinforced management. Nonetheless, in comparison with modern Japanese works, the British span of management is larger, the ratio being 1:82 at the Scunthorpe works and 1:62 on average in Kimitsu (excluding workers in in-works sub-contracting firms).

When compared with counterparts in Japan, the percentage of employment at Scunthorpe is smaller at the position of director (in Kimitsu 13.7 per cent), larger at the blue card grade (in Kimitsu 10.5

per cent), and almost the same at red card level (in Kimitsu 75.0 per cent) (Table 6.2). Thus we may assume that the responsibility and authority of directors are greater at Scunthorpe than in the typical Japanese steelworks and the division of functions at the blue card level is more advanced than in Japan (in Kimitsu, positions at this level are occupied by deputy departmental managers and special deputy managers, who are ranked intermediately between departmental head and section chief). Thus a sharper hierarchy at a higher level of management exists at Scunthorpe than in Kimitsu.

Managerial and non-managerial employees

Table 9.5 shows composition by managerial and non-managerial employees for the personnel department and for the blast furnace, LD converter and plate mill plants. The percentage of managerial employees in the personnel department is as high as 68.3 per cent and in breakdown the percentages of middle management and supervisors are both predominantly high when put together, accounting for 87.5 per cent of the total managerial employees of the personnel department. Middle management can be described as professional staff employees working in administrative functions. Supervisors are all leaders of the training and development section. Secretaries and clerks who are non-managerial employees are allocated to management and middle management employees as individual or common use staff under their supervision. Managers of the personnel department are divided into senior management and management ranks, and management-ranked personnel supervise and direct many middle management workers. In the personnel department, those who are in charge of the actual administrative function of that organisation are middle management personnel, who account for the greatest number of managerial employees.

The blast furnace, LD converter and plate mill have the highest proportion of managerial employees. This order is mainly the result of the difference in the percentage of foremen. The number of workers per foreman is 6.2 per cent at the blast furnace plant, 8.2 at the LD converter plant and 10.8 at the plate mill plant, thus showing that the span of management of a foreman is smallest at the blast furnace. The ratio of foremen to workers is thought to be determined by various factors, but some of the most important are the technical condition of each plant and the nature of industrial relations at the workshop level. The ratios in the early half of the 1970s were about eleven persons for the blast furnace plant and about fifty for the LD converter plant, thus showing

Table 9.5 Composition of managerial and non-managerial employees by direct and indirect departments, Scunthorpe works (numbers and per cent)

Composition	*Indirect department*				*Direct department*			
	Personnel department		*Blast furnace plant*		*LD converter plant*		*Plate mill plant*	
	(Number)	(%)	(Number)	(%)	(Number)	(%)	(Number)	(%)
Managerial employees								
Senior management (blue card)	1	1.2	1	0.4	1	0.3	1	0.2
Management (red card)	6	7.3	1	0.4	1	0.3	2	0.3
Middle management	25	30.5	8	3.4	9	3.0	16	2.5
Supervisory employees								
Supervisors	24	29.3	—	—	—	—	—	—
General foreman	—	—	4	1.7	1	0.3	—	—
Foreman	—	—	30	12.9	31	10.4	48	7.5
Total of managerial employees	56	68.3	44	18.8	43	14.4	67	10.5
Non-managerial employees								
Secretary/clerk	17	20.7	2	0.9	1	0.3	53	8.3
Workers	9	11.0	186	80.2	254	85.2	520	81.3
Total of non-managerial employees	26	31.7	188	80.1	255	85.6	573	89.5
Total	82	100.0	232	100.0	298	100.0	640	100.0

Source: Scunthorpe works (BSC): Interview

that the span of supervision was greatly reduced in the process of plant and equipment modernisation and rationalisation and that in consequence labour management has been reinforced.

The function of middle management at plants is different from that of the personnel department; it is to supervise and direct foremen, who are their subordinate rank. In other words, they are responsible for line function and in terms of titles the majority of them are section managers, assistant managers, shift managers and section engineers. Those who are in charge of the supervision and direction of middle management are those managers and engineers in the position of management (red card). Of the managerial employees at the plants, the foremen are most numerous. Although the span of the foreman's supervision and direction has been reduced, it is exercised over the workers who are in charge of actual operating tasks. In this sense foremen are performing a far more important managerial function than the supervisors of the personnel department, who do not immediately supervise clerks and secretaries.

When we compare the blast furnace plant and the LD converter plant of the Scunthorpe works with plants of a similar size (size of equipment and production capacity) in Japan (Kure works, Nisshin Steel Co. Ltd),[8] the number of levels in hierarchy from works manager to foreman is larger at Scunthorpe, being five in contrast to three in Japan. The span of supervision of the foreman is smaller at Scunthorpe, at 6.2 people at the blast furnace plant and 8.2 at the LD converter plant, while in Japan they are 13.6 and 13.2 respectively. However, while Scunthorpe has no subordinate managerial rank immediately below the foreman, one exists in Japan. The span of supervision of the group leader immediately below the foreman is thereby reduced to 3.2 and 4.3 people respectively. Thus, while the number of levels in hierarchy above the foreman is larger at the Scunthorpe works than in Japan, the number of the lowest units to be supervised and managed is also larger at Scunthorpe.

Workplace management

The work group at the workshop level is composed of workers who belong to different job promotion lines in the plant. The composition of the work group is here analysed for both blast furnace and LD converter plants at Scunthorpe.

The blast furnace operated by the blast furnace operating team was built in 1954 with a hearth diameter of 9.4 meters. It has one tap hole and is operated in a three-hour cast cycle. The charging of raw materials is all computer-controlled. The workforce composition of the blast furnace work team is shown in Table 9.6. This team carries out three different kinds of jobs required for blast furnace operation, for hob work, operation work and crane work. Hob work is carried out by a keeper and four helpers, operation work by a stove minder and charge attendant and crane work by one crane driver. The keeper in charge of tapping at the time (1984) was quite old and his length of service relatively long. He was responsible for the job requiring the most experience and skill in the work team. Yet his wage was not high enough to correspond to the differences of age and length of service. The other workers were mostly in their late thirties, with a length of service of around ten years. The majority of them entered the works in their late twenties as mid-career employees. In terms of length of service, by reference to Japan, they have already spent far longer in the job than is needed to acquire skill by experience.

Ordinary workers, excluding foremen, belong either to one of the two job promotion lines or a static job group which exists in the blast furnace plant (see Table 9.7). The two promotion lines are used to

Table 9.6 Workforce composition of blast furnace work team (Scunthorpe works, 1984)

Workforce	*Age*	*Length of service (yrs)*	*School career*	*Wage (£/week)*
Foreman	30–35	(min.) 5	Various	n.a.
Keeper	45–55 (52)	(min.) 10 (24)	Secondary school	167.38
Hob work				
1st helper	18–55 (45)	0–10 (17)	Secondary school	149.75
2nd helper	18–55 (38)	0–10 (9)	Secondary school	149.75
3rd helper	18–55 (38)	0–10 (9)	Secondary school	143.34
4th helper	18–55 (34)	0–10 (7)	Secondary school	143.34
Stoveminder	18–55 (38)	0–10 (9)	Secondary school	149.75
Crane driver	18–55 (34)	0–10 (7)	Secondary school	132.14
Charging attendant	18–55 (38)	0–10 (9)	Secondary school	154.56

Source: Scunthorpe works (BSC): Interview
Notes: (a) The age for finishing secondary school is 16 years old
(b) ( ) is the actual figures for 1981

Table 9.7 Blast furnace promotion lines and wages (Scunthorpe works, 1984)

Furnace – front side	(£/Week)	*Furnace – back side*	£(/Week)	*Static jobs*	(£/Week)
Keeper	167.38	Charging attendant	154.56	Stoveminder	149.75
Deputy keeper	159.37	Material dist. controller	149.75	OH crane driver	132.14
1st helpers	149.75	Coke wharf attendant	120.93	C/H operational	143.34
2nd helpers	149.75				
3rd helpers	143.34				
4th helpers	143.34				

Source: Scunthorpe works (BSC): Interview

promote workers by seniority based upon length of service.[9] At first, young workers and mid-career workers who enter the Corporation are pooled as apprentices and as soon as there is a vacancy somewhere on the line, the workers below that vacated job will move up, allowing someone from the pool to take up the bottom job on that line. Once in on that line, they move up the ladder, successively moving to jobs with higher wages. Static jobs do not form a promotion line and therefore those on these jobs do not change.

The blast furnace work team is composed of eight workers under one foreman who come from two different promotion lines and the static job group. The average age of a foreman is rather young and the minimum length of service required for this job is shorter than that of the keeper. The school careers of foremen are various, but to attain this position one must finish the NEBSS course[10] and have a qualification from it. In addition to this, young university graduates are encouraged to become foremen in order to increase their workshop knowledge and experience. Thus people relatively younger than ordinary workers and with longer academic careers may be chosen for foremen following public advertisements in the plant and sometimes they are recruited from other plants as well.

The LD converter work team is composed of eight workers under one foreman. These workers belong to two of the four promotion lines and the static jobs which are in the LD converter plant. The trends in age, length of service and school career are similar to the workers of the blast furnace work team in Table 9.6. Those of the foreman are also similar to those of the foreman in the blast furnace work team. As to the level of wages, the highest wage in the LD converter plant is higher than that of the blast furnace plant and the average wage is also higher for the LD converter plant. This difference is thought to reflect the agreement between management and unions to set wages higher for jobs in which related equipment has a larger value in terms of equipment investment. In addition, the fact that the LD converter is newer than the blast furnace means that it is easier to achieve redeployment to the new equipment if the level of wages is equal to or higher than that of the employee's former job. This is reflected as well in the higher wage level of the jobs in the continuous casting plant, which was built later than the LD converter plant.

The relationship among workers within the same work team is not like that of Japan, where it is based upon age and length of service with the foreman at the top. In Scunthorpe the workers are not only interested in their jobs and their wage rates, but also in other workers' jobs and wage rates. Therefore the wage rate (determined and changed through negotiation between management and the unions) may influence the opinions of the workers and create conflicts between workers as to the evaluation of their jobs. As to the relationship between foremen and ordinary workers, the foreman does not hold the functions of labour management and promotion control that are held by his Japanese counterpart, because workers are controlled by the principle of a seniority system of the union, and also qualifications and the methods of selection are different from those of Japan.[11] This aspect is reflected in the way that no position below foreman exists as it does in Japan.

Work organisation at the workplace is different from that in Japan. For example, at the blast furnace plant the jobs of hob work and that of the operation room are combined in the same work team, as are the jobs of the operation room and crane driver at the LD converter plant.[12] Such differences in work organisation are derived from concrete differences in individual equipment as well as the overall equipment situation of the works, such as the number of blast furnaces, the number of tap holes, the number of LD converters and whether continuous casters are installed or not. In addition, they are thought to derive from differences in industrial relations at the workplace, as seen in the seniority system and the job specification of the foreman.

CONCLUSION

Modernisation in plant and equipment affected the Scunthorpe workforce in much the same way as it did Japanese workforces, in that operatives decreased while the number of workers involved in engineering and maintenance increased. At Scunthorpe, however, it also acted in the sphere of works organisation, shifting relative importance from direct to indirect departments, implying a larger decentralisation of managerial functions than occurred in Japan.

Other points to note are that sub-contracting is limited and the division of labour at the works level is organised around the four main occupational classifications and also individual jobs. Trade union influence is visible in the system of seniority-based promotion as well as in job evaluation procedures. Foremen are not part of the labour management structure and have a correspondingly lower work status than in Japan.

The changes discussed above appear to fulfil the requirements of the Scunthorpe works to maintain a condition of contrived compatibility in management.

10 Management and labour at head office (UK)

This chapter provides us with important evidence that the corporate structure and the function of head office of the BSC reflects an advanced stage of development in corporate structure. The frequent change in corporate organisation is to a large extent due to political and legislative influences. A quantitative increase in managerial personnel may be identified as a phenomenon in common with Japan; Britain manifests thereby an increase in contrived compatibility, while Japan shows organic compatibility in management. The employee director scheme which developed in BSC is another indication of contrived compatibility in the area of industrial relations, but one that did not appear in Japan.

CORPORATE ORGANISATION

A new corporate organisation of the BSC was established in March 1970, based upon the *Third Report on Organisation*,[1] announced in December 1969. The main characteristic of this organisation lies in the establishment of a system of product divisions, each operating as a profit centre with a greater control than before from head office. This was the organisation that aimed to create a crude steel capacity of 35 million tonnes announced later in the *Ten Year Development Strategy* of February 1973 (DTI 1973: Cmnd 5226).

In April 1976, this form of organisation was discontinued and another adopted. This variant, while retaining profit centres, partially changed the iron and steel sector into geographical cost centres (the five manufacturing divisions) with an establishment of four product units in charge of the sales and production planning. The product units were under the direct control of head office, thus increasing their control over sales and production planning. In September 1980, there was yet another organisational overhaul, again reflecting the sharp decline in crude steel production. The new order of things was established to deal with the fact that increases in demand had not taken place as had been expected in *Road to Viability* (DTI 1978: Cmnd 7149),[2] *The 15 Million Tonnes Strategy* (BSC 1979) and the *Survival Plan* (BSC 1980).

The corporate organisation introduced in April 1970 is shown in Figure 10.1. This organisation consists of the board, chairman/chief executive, functional organisations of head office and line organisation (product divisions). The characteristic of this organisation is the adoption of a product division organisation which functioned as a product-based profit centre by replacing the geographical group organisation that had been in place since nationalisation. This aimed to eliminate a problem that had existed in the old organisation, under which most products fell within more than one group, and to achieve the maximum long-term return on its capital investment[3] by rationalising sales, employing its plants to the maximum benefit and planning its capital investment programme (BSC 1969b: 7).

The corporate organisation implemented in April 1976 is shown in Figure 10.2. This organisation consists of the board, chairman, deputy chairman/chief executive, functional organisations of head office and line organisation (manufacturing divisions and profit centre). The characteristics of this management organisation were that the chairman was no longer chief executive and the iron and steel sector, which accounted for about 80 per cent of the total sales of the Corporation, was changed into geographical manufacturing divisions to function as the cost centre. The fact that the chairman was no longer chief executive suggests that the organisation was revised so as to make it possible for the chairman to concentrate exclusively on board matters. In addition, the change of the iron and steel sector into manufacturing divisions suggests that the necessity to increase head office control over production and sales arose as a result of the concentration of production in the major large-scale works through plant and equipment modernisation and rationalisation. This had become technically possible due to the development of computerisation (BSC 1975).

The allocation of orders to the manufacturing divisions/works in consideration of the overall production and commercial planning of head office was carried out through five newly-established product units, which were under the direct control of head office (BSC 1975: 8–9). Orders were allocated to plants to the best financial advantage of the Corporation, consistent with plant and material constraints, customer location, the level of trade, labour and industrial relations considerations, and other factors (BSC 1975: 9). The commercial headquarters was established near each product unit and here integra-

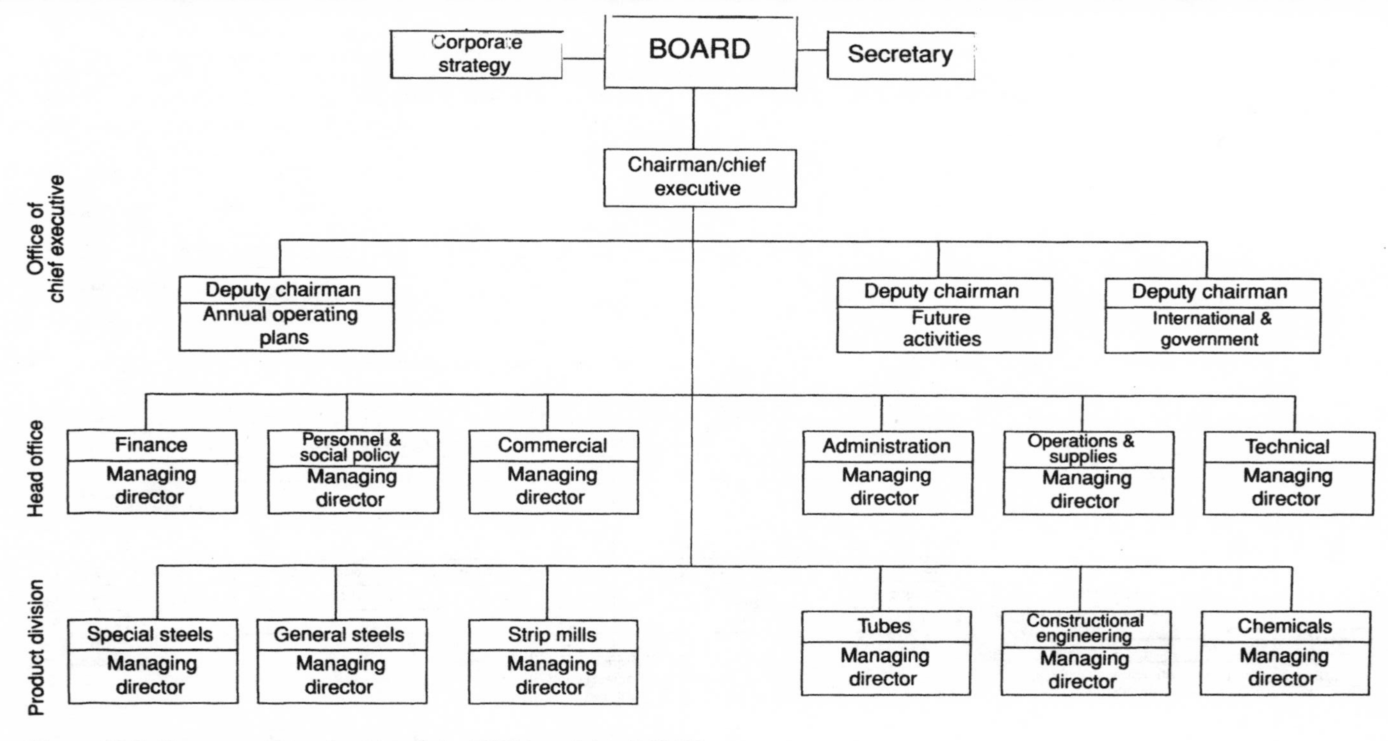

Figure 10.1 Corporate Organisation of the BSC (as of April 1970)

Source: The chart was produced with slight modifications from BSC 1969: 20–1 Crown Copyright is reproduced with the Persmission of the Controller of HMSO.

Note: Chairman indicates chairman of the board and deputy chairman indicates deputy chairman of the board. The board is composed of chairman, deputy chairman and executive and non-executive board members.

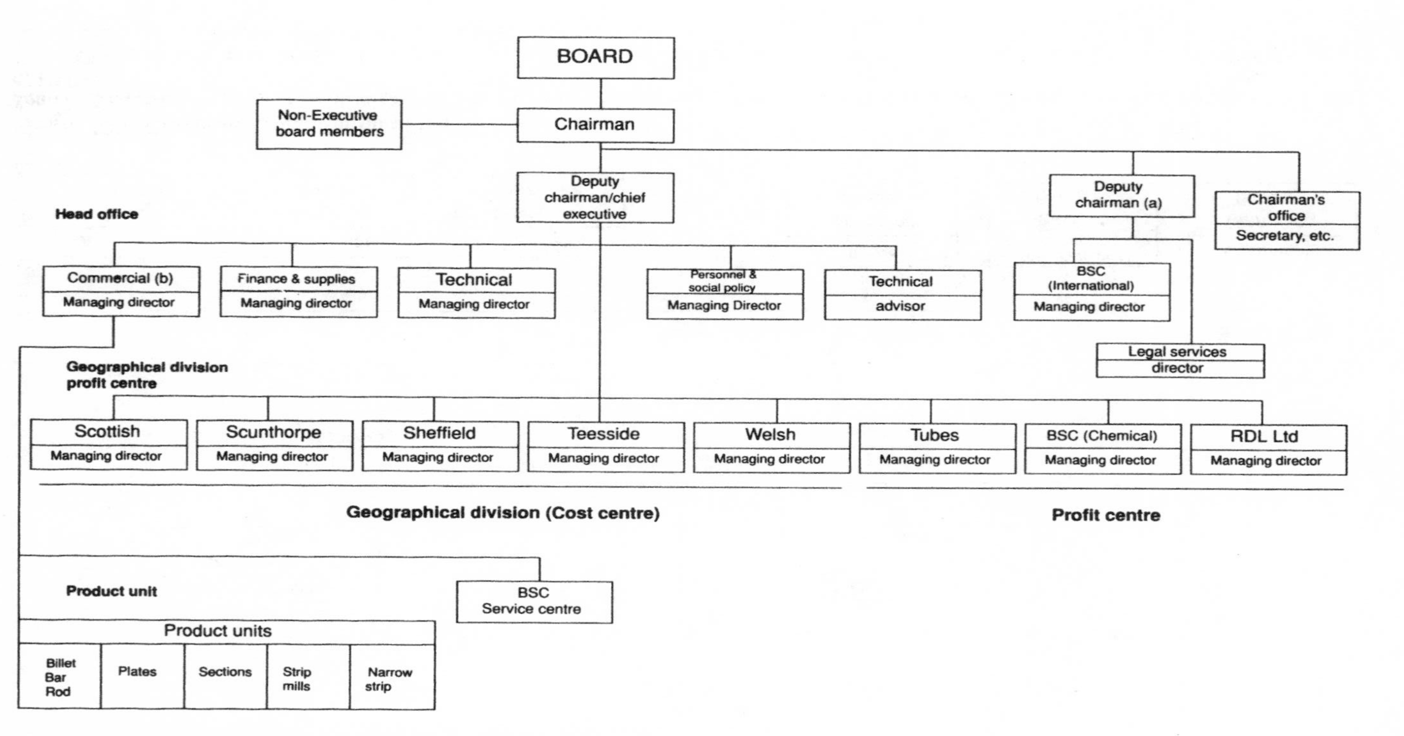

Figure 10.2 Management organisation of the BSC (April, 1977)

Source: The chart was produced with slight modifications from the BSC, *Annual Reports and Accounts*, 1976/77

Notes: (a) Chairs the two board committees (finance committee and audit committee)
(b) Present holder is also a board member
(c) Serves chairman and both deputy chairmen collectively

tion of the manufacturing and sales activities was carried out in addition to the considerations of cost, price, the competitive environment and profitability of products (BSC 1975: 9). Such changes in organisation were expected to strengthen competitive power, particularly in the European Economic Community (BSC 1975: 5).

Last, the management organisation which was implemented in September 1980 is shown in Figure 10.3. This organisation consists of the board, chairman/chief executive, deputy chairman/chief operating officer, functional organisations of the head office, and line organisation (operating groups and businesses). The characteristics of this management organisation are first, that the chairman has returned to take the position of chief executive, while the deputy chairman has assumed responsibility for substantial operating functions, thus assisting the chairman who is not from an iron and steel background; second, the existing divisional organisation has been abandoned and the line organised into three operating groups within which many product-based businesses were established. The change to a series of discrete, decentralised, product-based businesses, each acting as a profit centre, was due to the understanding that in the old organisation head office was thought to be able to control business activities in detail and thus had grown in size in order to increase the function of detailed reporting and monitoring and the provision of comprehensive central services. However, this arrangement no longer fitted into the reduced operational size under the competitive pressures, and was in fact disadvantageous (BSC 1980). The full-scale business reorganisation was expected to stimulate each individual business activity and develop and rationalise comprehensive management by making each business independent (BSC 1980). In addition, it was expected to prepare businesses with good operational records for privatisation under the reins of the Conservative government. In fact, it is reported that the total bulk value of assets disposed of between 1980–4 was £274 million. Redpath Dorman Long Limited (an engineering business), which had a good business record, was also disposed of in 1982 (BSC, *Annual Report and Accounts* 1981/82).

The revision of corporate organisation (revised three times since 1970), shows the result of decisions by top management of the large-scale iron and steel enterprise to deal with the changing corporate environment, on the basis of plant and equipment modernisation and rationalisation. The characteristics recognised as common to all three organisations are:

1 there exists a board as a supreme decision-making organ,
2 there exists an office of chief executive which consists of chairperson and deputy chairperson/s

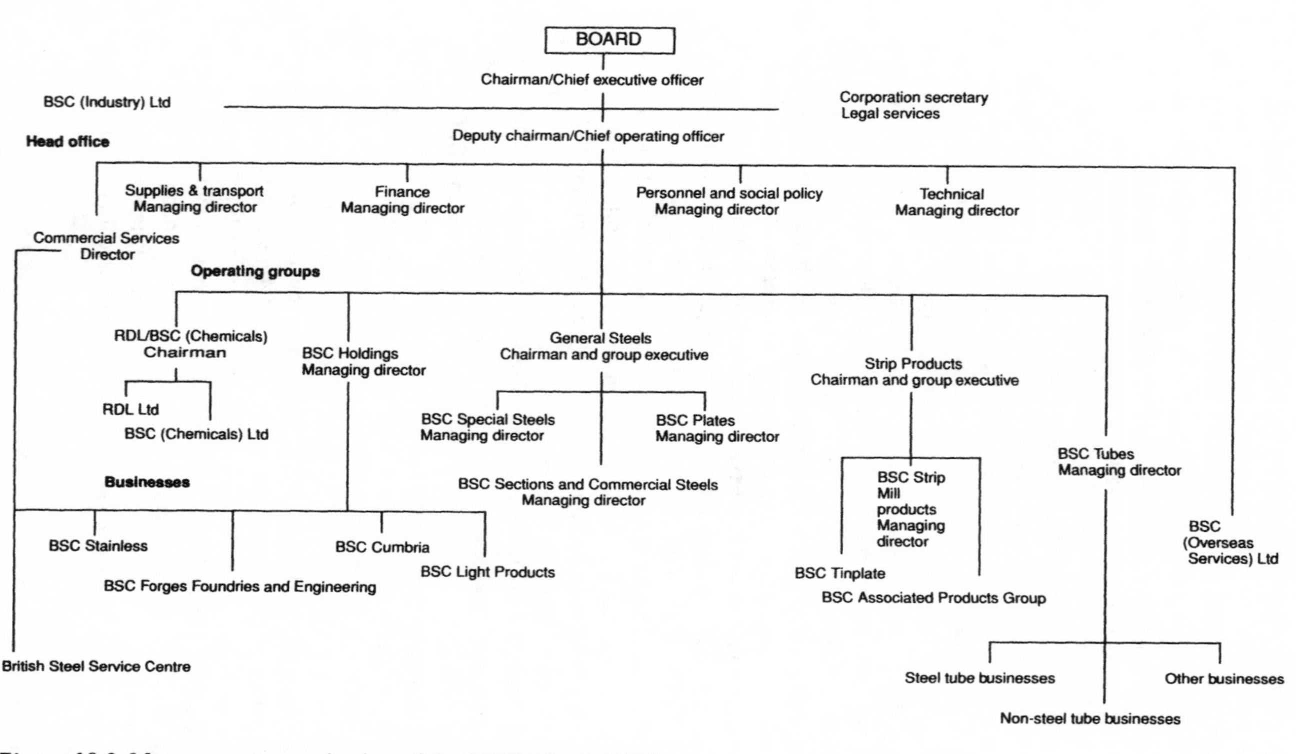

Figure 10.3 Management organisation of the BSC (March 1981)
Source: The chart was reproduced with slight modifications from BSC, *Annual Report and Accounts*, 1980/81

3 the range of functional organisation of head office is rather small, having four to six functions,
4 the line organisation always has a profit centre (business) organisation (partly or wholly).

A major difference can be recognised in the structure of organisation between the period when equipment modernisation and rationalisation was carried out with a forecast of demand increase (1970s), and the period when the situation of demand decrease and weakening of competitive power was dealt with (1980s). The 1970s saw the size of head office enlarged and the management function of the line organisation accordingly partly transferred to head office. It was intended, under the expectation of increased production, that head office, using computers as its technical basis, would control centrally several large-scale integrated works which would make the best use of technical scale merits, thus constituting a sign of partial convergence to Japan. The embodiment of this attempt is reflected in the management organisation of the period 1976–80, which made the works perfect cost centres. On the other hand, in the period from 1979 into the 1980s, when demand decreased and competitive power weakened rapidly, the structure of organisation was changed to a business system attempting to restore financial and competitive strength by sub-dividing the line organisation, as the conditions needed to exploit the scale merit of the large-scale integrated works had been lost.

HEAD OFFICE ORGANISATION

The functions of head office can be divided broadly into two. One is the function which concerns corporate management as a whole and the other is the service function to the line organisation. The former is the function which has been maintained at a constant scale in spite of changes in the corporate organisation and the latter is the function which has increased or decreased in accordance with the changes in the corporate organisation.

The general functions of head office in the management organisation in the period 1970–6 were as follows:

1 finance (the development of standard costing systems, the control of capital expenditure, the financial assessment of plans and capital expenditure proposals, the co-ordination and financial monitoring of annual operating plans, banking, and taxation);
2 development planning;
3 raw materials procurement policy;

4 the acquisition and disposal of interests;
5 the overall determination of personnel and social policy, including industrial relations policy, management development, salary structure and pensions;
6 regional policy (consideration of interests in Scotland and Wales);
7 overseas interests;
8 the overall determination of commercial policy (marketing policy, annual sales plan, prices and sales policy in relation to both home and export markets).

(BSC 1969)

The service functions rendered by head office were as follows:

1 legal services;
2 central purchasing of some materials;
3 economic forecasting and other economic services, and certain management services.

(BSC 1969)

In the post-1980 organisation, the reduced head office still retains two basic functions. One function, concerning the whole Corporation, can be broken down as:

1 directing and guiding the businesses on essential matters of central policies and also matters between the Corporation and the government, ECSC and the National Trade Union movement;
2 ensuring that businesses have plans and strategies to meet financial objectives;
3 raising and allocating resources to secure overall corporate objectives;
4 assessing the Corporation's overall performance; and
5 providing high-quality advice to the chairman and deputy chairman on short-term problems, long-term trends and development activities.

The other function is the minimum service function, and can be broken down into:

1 the service function for which it is uneconomic to allow separate provision in each business, or which requires co-ordination;
2 the service which can exploit the scale merit of the economy; and
3 services on legal matters, industrial relations, statistics and pensions.[4]

Table 10.1 shows the number of employees classified by the functional organisation of head office and its related offices between 1977 and 1982. Concerning head office and its related organisations, the organisation with the largest number of employees in both periods is the technical organisation, which accounts for nearly 70 per cent. The reason for this high percentage in the technical organisation is that R & D offices are included in head office organisation. Besides the technical organisation, organisations which have a relatively large number of employees are supply and transportation, personnel and social policy, and finance. Those with a rather small number of employees are internal audit and commerce. The decrease rate of employees is largest in commerce, followed by administration, finance, personnel and social policy, and the board and related organisations. A significant reduction in the number of employees in these organisations suggests that a large part of the head office functions which belonged to the three organisations were delegated to the businesses under the 1980 restructuring. In particular the service function of each organisation was reduced and delegated to the line. That is to say that the old organisations in the 1970s had tended to have an active interface politically and industrially, and businesses were treated as secondary in the environment of strong central functions, while the new organisation implemented in September 1980 tended to have an interface at a minimum level and transferred problems and questions of business management, functions of service and monitoring to operating groups and businesses.[5]

Table 10.1 Composition of employees at BSC head office (all locations), by organisations

Organisation	*1977*		*1982*		*% change 1982/1977*
Board and related	138	(4.1)	88	(3.5)	−36.2
Commercial	120	(3.6)	35	(1.4)	−70.8
Finance	225	(6.7)	132	(5.2)	−41.3
Internal audit	67	(2.0)	69	(2.7)	3.0
Supplies and transport	239	(7.1)	200	(7.9)	−16.3
Personnel and social policy	231	(6.8)	146	(5.7)	−36.8
Administration	202	(6.0)	115	(4.5)	−43.1
Technical	2,157	(63.8)	1,759	(69.1)	−18.5
Total	3,379	(100.0)	2,544	(100.0)	−24.7

Source: BSC: Interview
Note: (a) The figure for 1977 is as of June and 1982 as of March
(b) the number in percentage

Through the above examination of head office organisation, the role of this organisation can be classified into two in every period. Important functions concerning the whole Corporation have been constantly maintained at a certain level. The monitoring and service functions to the line either increased or decreased, depending on the change in the management organisation. The commercial organisation was enlarged when the role of head office was strengthened, but even at that time was relatively small and has been further reduced since 1980. This suggests that the majority of the commercial functions of the Corporation have always been delegated to establishments or related organisations, even including the period when the role of head office was strengthened. If we exclude technical organisation, the size of head office is extremely small when compared to the size of the Corporation as a whole. This reflects the federal control system which allows a comparatively high level of independence to establishments in terms of delegation of management functions.

Managerial employees of the Corporation

The impact of plant and equipment modernisation and rationalisation upon the composition of managerial employees from the first half of the

Table 10.2 Composition of managerial employees at the Corporation level (iron and steel sector only)

Employees	*1972 (October–December) (Number)*	*(%)*	*1979 (October–December) (Number)*	*(%)*	*% Change 1979/1972*
Top management					
Board members	12	0.007	20	0.01	66.7
Managing directors	12	0.007	13	0.01	8.3
Senior management					
Managers	114	0.06	689	0.5	504.4
Management in special functions	6,063	3.5	6,629	4.9	9.3
Scientists and technologists	1,721	1.0	1,399	1.0	−18.7
Functional specialists	4,412	2.5	3,791	2.8	−14.1
Supervisory					
Supervisors	2,358	1.3	2,195	1.6	−6.9
Foremen	7,270	4.1	6,203	4.6	−14.7
Clerical employees	25,360	14.5	20,273	15.0	−20.1
Manual employees	128,158	73.0	93,615	69.4	−27.0
Total	175,480	100.0	134,827	100.0	−23.2

Sources: BSC: Interview; BSC, *Annual Report and Accounts*, 1971/72, 1978/79.

Notes: (a) The number of managing directors is as of the end of March of each financial year. In 1979, the managing director of BSC (Chemicals) Ltd position was vacant, therefore not included in the number for that year

(b) In the position of supervisors clerical supervisors (male) and clerical supervisors and technicians (female) are included

1970s to the early part of the 1980s is shown in Table 10.2. While the total amount of employment decreased by 23.2 per cent on average, top and senior management increased. The fall in the number of functional specialists and supervisory employees is relatively small, while that of scientists and technologists and clerical employees is quite substantial. The greatest decrease occurred among manual workers.[6]

The ratio between blue- and white-collar employees shifted in the same period from 73.0 : 27.0 per cent to 69.4 : 30.6 per cent, thus resulting in the total of white-collar employees exceeding 30 per cent. The percentage of white-collar employees other than non-managerial clerical workers rose from 12.5 to 15.6 per cent, thus shifting to a composition appropriate to increased managerial control in the overall Corporation. In top management, the number of board members shot up in 1979 due to the participation of employees on the board, while in senior management a remarkable increase of managers and an accompanying increase in those holding positions in special functions management is noted. The percentage of those in management in special functions is quite high compared to the total of managerial employees, implying the importance of their role both in quantity and quality. In contrast to the large decrease in the number of manual workers, that of supervisory employees is not so great, resulting in a decrease in span of management by supervisory employees from 24.1 to 21.7 persons. In consequence this has created a shift of composition which enables the provision of small-span management at workshop level. Thus the change in composition of managerial employees suggests two things: an increase in the functions of senior management and supervisory employees, and organisational inertia to build up the managerial hierarchy.

Employees by head office and other establishments

The change in the number and composition of employees at head office (all locations) is shown in Table 10.3. In the period 1972–9, when plant and equipment modernisation and rationalisation were most substantial and the control of head office over other establishments was strengthened, despite a big fall in the number of employees in the whole Corporation, the number of employees in head office and its related organisations more than doubled, thus raising the percentage in composition from 1.2 per cent to 3.5 per cent. But in the period 1979–81, the number of employees allocated to head office (all locations) also decreased in line with the sudden and drastic drop in the total of Corporation employees. In terms of percentage, however, as the overall

Table 10.3 Number and composition of BSC employees by head office (all locations) and other establishments

	1972		*1979*		*1981*	
	(Number)	*(%)*	*(Number)*	*(%)*	*(Number)*	*(%)*
Head office						
(all locations)	2,380	1.2	5,108	3.5	4,325	4.4
(laboratories)	(818)	(0.4)	(1,634)	(1.1)	(1,431)	(1.4)
Other establishments	191,382	98.8	142,158	96.5	94,529	95.6
Total	193,762	100.0	147,266	100.0	98,854	100.0

Source: BSC: Interview
Notes: (a) The figures for 1972 and 1979 are as of December and for 1981 as of March
(b) The number of employees is limited only to the iron and steel sector of the Corporation
(c) Employees at laboratories in 1972 are not organisationally included in head office (all locations)

loss of employees was so great, the percentage of employees in head office and its related organisations continued to rise even in that period. The change in the number and composition in research and development offices shows a similar trend to that of head office. In terms of composition by male and female, in the period 1972–81 the female percentage declined slightly from 33.9 to 32.6 per cent.[7]

The above examination of the number and composition of employees suggests that in the period 1972–9, when the control of head office increased, the number of employees in head office and its related organisations rose significantly, due to the fact that many organisations such as laboratories came to be included in this category in addition to the increase of employees in head office itself, which is located in London.[8] The percentage of employees in head office and its related organisations continued to rise even after the implementation of the new organisation which caused a reduction in head office control, accounting for 4.4 per cent in 1981. This shows that plant and equipment modernisation and rationalisation and the consequent reorganisations of management structure have provided a condition for a relative enlargement in the size of head office. The fact that the composition rate of total head office employees increased while that of the female composition decreased suggests the progress of rationalisation in office work by the use of computers.

Table 10.4 Composition of BSC managerial employees at head office (London)

		1972		*1979*		*1982*	
		Number	*(%)*	*Number*	*(%)*	*Number*	*(%)*
Top Management							
Board	Executive	6	0.4	5	0.3	4	0.3
member	Non-executive	6	0.4	15	0.9	11	0.9
Managing	Function	6	0.4	5	0.3	5	0.4
directors	Line	6	0.4	9	0.5	8	0.7
Managerial employees							
Senior management				630	36.8	425	36.7
					(54.8)		(51.2)
Management		950	59.9	170	9.9	165	14.2
					(14.8)		(19.9)
Middle management				350	20.4	240	20.7
					(30.4)		(28.9)
Sub-total		950	59.9	1,150	67.1	830	71.6
					(100.0)		(100.0)
Non-managerial employees		612	38.6	530	30.9	300	25.9
Total		1,586	100.0	1,174	100.0	1,158	100.0

Sources: BSC *Annual Report and Accounts*, 1971/72, 1978/79, 1981/82; BSC: Interview

Notes: (a) One line managing director position in 1979 was vacant

(b) Two members included in the category of line managing directors in 1982 are the chairmen of the operating groups

Managerial employees at head office

Table 10.4 shows the composition of managerial employees at head office (London) classified by top management, managerial employees and non-managerial employees. The board members, who constitute top management, are appointed by the minister after consultation with the chairman of the Corporation and they become members of the board, which is the supreme decision-making organisation. There are full- and part-time board members. In 1972, when the management organisation was under the profit centre system, there were six full-time and six part-time members, making twelve in all. In 1979, when the Corporation was under the geographical division system, the number of full-time members was five while that of part-time members was fifteen, making twenty seats on the board. By 1982, when the Corporation was under the group and business system prior to privatisation in 1988, the number of board members had shrunk to fifteen, eleven of whom were part-time. Thus the number of full-time members has been declining since 1972 while a big jump occurred in part-time members in the period 1972–9, followed by a decrease. This is partly due to the introduction of an employee director scheme at board level, which brought six members to the board in 1978–81 and four in 1982. It has been decided that some of the managing directors in charge of functions are qualified to become full-time board members, but those in charge of line responsibility are not so qualified because they are expected to devote themselves to the line responsibility (BSC 1969b).

The chief executive, who is either chairman or deputy chairman, is delegated all executive authority by the board. He then delegates his authority and responsibility to the managing directors of both functions and the line. Namely, managing directors in charge of functions will advise on and recommend policies, programmes and procedures on all matters relating to their functions (BSC 1969b). The managing directors in charge of line activities delegate further their authority and responsibility to the directors in charge of functions and line responsibility of that line who constitute top management (BSC 1969b). In 1972, when the Corporation was under the profit centre organisation, there were six functional managing directors and six line managing directors. In 1979, when the Corporation was under the geographical division organisation, there were fourteen directors in all, five and nine respectively in function and line mangement. Finally, in 1982, when the Corporation was under the group and business organisation, there were five and eight (two of the latter were chairmen of the operating groups). Thus, while there has been almost no change in the number of functional managing directors, line managing directors increased slightly in 1979 when compared to 1972 and in 1982 in comparison to 1979. In 1982 two of the eight top management members responsible for the line were chairmen of operating groups, whose status is higher than that of the managing directors, and in addition one of the two chairmen was concurrently a full-time board member, thus creating a further hierarchy in the executive positions of the top management.

Below the top management there are managerial employees classified into senior management, management and middle management. The authority and responsibility delegated to a managing director is sub-divided horizontally and delegated to a small number of directors and managers (senior management) in that organisation (for further details see the following section). The number of senior management is said to have increased in the period 1972–9 but it decreased drastically in the period 1979–82. The rate of senior management to total managerial employees is quite high, exceeding 50 per cent. The authority and responsibility delegated to senior management is then sub-divided into a hierarchy composed of management and middle management. Management numbers are also said to have increased in the period 1972–9, but decreased slightly towards 1982. The proportion of management to total managerial employees is at or above the 10 per cent level, but rose to nearly 20 per cent in 1982. The middle management in the head office managerial grade, who are in the lower professional staff bracket with no managerial executive authority, can also be seen to have increased in the same period, but decreased greatly in the period 1979–82. The proportion of middle management among the total of managerial employees is about 30 per cent. The percentage of managerial employees in the composition of total employees at Head Office continued to rise from 59.9 per cent in 1972 to 67.1 per cent in 1979 and then to 71.7 per cent in 1982, which resulted in a relative expansion of the managerial function of head office as well as the function to manage managerial personnel at head office.

We also find that the percentage of top management continued to rise from 1.6 per cent in 1972 to 2.0 per cent in 1979 and to 2.3 per cent in 1982, thus enlarging the relative size of top management. This enlargement is caused by the rise in percentage of part-time board members and line managing directors. This suggests that while in this group subdivision of authority and responsibility and hierarchy have developed, the relative authority and responsibility of the full-time board members and the functional managing directors whose percentage remains constant or has declined were increased; in particular, after 1980 the four full-time board members (chairman, deputy chairman, functional managing director, chairman of an operating group) came to hold substantially increased authority and responsibility, acting as members of top management concurrently performing the functions of both decision-making and execution.

Although in 1982 the chief authority and responsibility were concentrated in the four full-time board members, in general, due to the method of appointment of the large number of part-time board members, the small number of functional managing directors among the full-time board members, and the non-appointment of line managing directors to the board, it is suggested that the board is organisationally independent of the line executive function and that important policies and strategies are affected by the composition of the board members at any one time, thus raising the possibility of a loss of continuity in the nature of policies and strategies. The number of functional managing directors is very small, remaining around five or six. The functional organisations which have existed throughout the period since 1970 are finance, personnel and social policy and technical and supply. The importance of the function of personnel and social policy has been consistently high, which suggests that equipment modernisation and rationalisation has always made a great impact upon the employment of a particular area and also upon industrial relations. Other functional organisations have changed their level of importance in accordance with the change of the whole management organisation. For example, the commercial organisation increased in importance, especially in the period when the control of head office was strengthened, but after management organisation changed to that of operating groups and businesses in 1980, it was downgraded even in terms of organisation. Within the group of managerial employees the most numerous are senior management, followed by middle management, who are functional specialists. In recent years the percentage of management has risen, thus suggesting the development of a division of managerial functions at this position. In addition, the percentage of senior management at head office is extremely high compared to that of other establishments, thus showing a high level of concentration of senior management at head office.

Comparison of managerial employees – head office and a selected establishment

The personnel and social policy headquarters are high in importance among head office functional organisations and consist of seven departments: health and safety, industrial relations, personnel development and services, pensions administration, pensions investment, information services and administration. The large and prominent industrial relations and personnel development and services departments are controlled by directors, but other organisations are controlled by managers (senior management). The industrial relations department has four sections controlled by managers and one principal industrial officer. These four sections are the industrial section, staff section, development section and manpower planning and social policy. There is one managerial employee under the manager of the industrial section, and three (the majority being management and middle management) under the manager of each staff section: development section, manpower planning section and social policy section. In the industrial relations department, in addition to these managerial employees there is one clerk, one senior secretary, and three secretaries who are not managerial employees. As to the classification of managerial employees in the industrial relations department by sex and academic career, those who are senior management are all male and university graduates, while those in management and middle management are females and non-university graduates. Thus we can recognise a relationship between managerial rank, gender and academic career.

In Table 10.5 there is a comparison of the composition of managerial employees with similar managerial functions between the industrial relations department of head office and that of the personnel department of the Scunthorpe works. The percentage of the total managerial

Table 10.5 Composition of BSC managerial employees by head office and a selected establishment

Employees	*Head office (industrial relations dept 1983)*		*Scunthorpe works (personnel dept 1984)*	
Managerial employees	(person)	(%)	(person)	(%)
Senior management	7	(33.3)	1	(1.2)
Management	5	(23.8)	6	(7.3)
Middle management	4	(19.0)	25	(30.5)
Supervisory			24	(29.3)
Sub-total	16	(76.2)	56	(68.3)
Non-managerial employees				
Secretarial, clerical	5	(23.8)	17	(20.7)
Manual workers			9	(11.0)
Sub-total	5	(23.8)	26	(31.7)
Total	21	(100.0)	82	(100.0)

Source: BSC: Interview
Note: One director is included in the number of senior management of the industrial relations department at head office

employees is considerably higher for the head office industrial relations department, accounting for 76.2 per cent as against 68.3 per cent for the personnel department of the Scunthorpe works. In terms of managerial ranks, there is no one at supervisory rank in the industrial relations department of head office, while those in the personnel department of the Scunthorpe works account for just under 30 per cent. In addition, among non-managerial employees there are no manual workers in the industrial relations department of head office, while those in the personnel department of the same works account for 11.0 per cent. In terms of the breakdown of managerial employees, the percentage of senior management and management is extremely high in the industrial relations department, accounting for 57.1 per cent, while that of the personnel department is as low as 8.5 per cent. But the percentage of middle management is relatively low (accounting for 19.0 per cent) in the industrial relations department, while it is relatively high in the personnel department of the same works, accounting for 30.5 per cent. Among non-managerial employees the percentage of secretaries and clerks in the industrial relations department is slightly higher than that of the personnel department of the same works.

The above comparative examination of managerial composition between the two departments, one in head office and the other in the works, suggests that in the industrial relations department that belongs to the personnel and social policy function of head office the horizontal division of managerial functions performed by senior management and the development of hierarchy in management and middle management have progressed. We can also observe a certain inter-relationship between the managerial ranks and academic career and gender. The industrial relations department has a much higher percentage of employees in the upper ranks of managerial function such as senior management and management compared with the personnel department of the works. The industrial relations department does not have any supervisory employees and manual workers, while its percentage of secretaries and clerks is higher than that of the personnel department of the works. This difference reflects a characteristic of the functional organisation of head office as a centre of control.

Employee directors and board members

Employee participation as non-executive members on the board at head office started in August 1978, but prior to this the BSC had, since nationalisation, promoted employee participation in management at establishment level as a part of industrial democracy.[9] As is shown in

Table 10.6 Number of BSC employee directors and main board members

	Directors of establishments (group, division, business)	Main board (non-executive members)	Total
1968	12	0	12
1970–3	16	0	16
1974–6	17	0	17
1977	The employee director scheme frozen from April 1977		
1978	14	6	20
1979	9	6	15
1980	7	6	13
1981	6	6	12
1982	6	4	10

Source: BSC: Interview
Note: The group system existed from April 1970 to March 1976, that of division from April 1976 to August 1980, and that of business from September 1980 to present

Table 10.6, employee participation in management can be classified into two levels, one at establishment level (board members of groups, divisions, businesses) and the other at head office level (non-executive board members). The number of directors at establishment level continued to increase from 1968 to the period 1974–6, reaching a peak of seventeen. It began to decrease, however, in 1978 and was down to six in 1982. The number of employee board members at head office remained at six during the period between 1978–81 but went down to four in 1982. In terms of the total number of employee directors and board members, the employee participation which began in 1968 reached its peak in 1978 and then began to decline and it was at its lowest level in 1982.

The employee participation scheme was promoted by the initiative of the Corporation from the start of nationalisation. It was thought that against the general trend of industrial relations in Britain and the rapid increase in unofficial strikes after 1970,[10] the Corporation, which had been formed through the integration of fourteen companies, could not carry out any standardisation of labour relations practices, reform of payment systems, or plant and equipment modernisation and rationalisation without the co-operation of its workers. Although the Corporation had a statutory obligation to develop joint consultation and collective bargaining with the trade unions, the organising committee, which had been formed to establish the Corporation, recognised the positive effects of employee participation for smoothing and stabilising industrial relations and proposed the scheme to the TUC (Bank and Jones 1977: 12–13). The employee directors and board members are not selected by means of election and thus they do not organisationally represent the trade unions. The TUC thought at first that the scheme was significant in so far as the employee directors would represent their unions, but in the end it was agreed that employee directors would relinquish their trade union position when appointed as directors of the group of the district to which they belonged (Bank and Jones 1977: 13).

As a result of the re-examination of the employee director scheme conducted from 1969 to 1971 by four independent academics (Bank and Jones 1977: 35–7, 63–8), it was decided in consultation with the Steel Committee that employee directors would be able to hold their union offices and maintain close links with the unions. In addition, Paragraph (vi) of the old job description was amended and two paragraphs were added in order to extend involvement in joint consultative meetings, advisory committees and working parties (Bank and Jones 1977: 110). But the method and procedure of selection and appointment was kept as it was before.

The percentages of employee directors and employee board members to the total in each period of the different management organisations, such as group, division and business organisations, has always been less than 35 per cent. For example, in 1968 it was 25 per cent in the Midland group, 19 per cent in the Northern and Tubes group, 20 per cent in the Scottish and Northwest group and 19 per cent in the South Wales group. In 1974, it was 33 per cent in the General Steel division, 28 per cent in the Strip Mills division, 25 per cent in the Special Steels division, 23 per cent in the Tubes division, 13 per cent in the Chemicals division and 7 per cent in Redpath Dorman Long Ltd. In 1979, it was 33 per cent on the board, 16 per cent in the Scottish division, 18 per cent in the Teesside division, 18 per cent in the Tubes division, 25 per cent in the Welsh division, 14 per cent in the Scunthorpe division and there was no participation in BSC Holdings, BSC (Chemicals) Ltd, Redpath Dorman Long Limited or BSC (International) Limited. Thus the percentage rate, which may be seen as a quantitative index for participation, began to decrease towards the end of the 1970s and declined further in the 1980s. The reason for this would seem to be that after the termination of membership in 1981 four out of six members were reappointed by the Board but two nominations by the ISTC were not accepted by the chairman; the ISTC executive committee then decided to withdraw from the board, resulting in the decrease of its number to four (Interview 1987). In addition, after 1978 directors at establishment level were not replaced when they retired, so that the number steadily decreased to as few as six in 1982.

As indicated above, the shift in the number of employee directors and board members is a reflection of industrial relations, and in particular reflects the strategy of the industrial relations management in the BSC. This strategy has been pursued positively as part of the managerial participation policy which began at the level of individual establishments as soon as nationalisation took place. But considerable difficulties arose when the *Survival Plan* was introduced in 1980 and the role of employee directors had to legitimise the drastic rationalisation proposed by the BSC, which led to a certain distancing effect with the unions and eventual curtailment of the scheme.

CONCLUSION

The corporate structure of BSC is based on the concept of decentralisation of corporate functions and during the period under study it increased the degree of this decentralisation while maintaining the basic control of head office. This can be regarded as a feature of a mature corporate structure, which emphasised profitability in each division. The quantitative increase of managerial staff at head office, on the other hand, seems to be an unnecessary expansion of the managerial hierarchy in the above context of decentralisation. Industrial democracy in the form of employee directorship did not seem to function as well as had been hoped, either for management or for unions, and began to decrease in importance at the stage of contrived compatibility, giving management the opportunity to reformulate industrial relations to its own advantage.

Part IV

Convergence and relative advance

11 Japan/UK comparison of management and labour

The objective of this chapter is to compare the changes in management and labour identified in Parts II and III. This will be carried out by a comparison of salient differences and similarities in the style of equipment modernisation and then in management and labour. We will also consider contributory factors for such differences and similarities, and see if such phenomena can be explained in terms of organic or contrived compatibility in management. The results of the comparison made in this chapter will reflect the dynamics between technology, managerial strategy and labour unions.

CORPORATE ENVIRONMENTS

The rapid and massive investment in the Japanese steel industry was due to macro-economic conditions characteristic of the stage of organic compatibility in management. These conditions are a faster growth in demand allowing new technology, strong domestic competition, consistent government industrial policy and the formulation of good industrial relations. As shown in Tables 5.1 and 8.1, the timing of capital investment in Japan and the UK was different; Japan invested in the 1960s, while the UK did not get around to it until the late 1970s. We can explain the macro conditions for the Japanese steel industry as follows:

1 Domestic demand increased greatly in this period due to large and steady growth and development in the construction, shipbuilding, automobile, machinery and petro-chemical industries (see Chapter 3, pp. 35–7).
2 There was a rapid, large increase in exports, especially to the United States. Conditions were favourable for this because the US steel industry was becoming relatively weak in competitiveness after

1960. The rapid increase of steel exports from Japan increased Japan's share of world steel production from 6 per cent in 1960 to 16 per cent in 1965 and 20 per cent in 1970.

3 Domestic competition among the five major corporations since 1960 has promoted capital investment in order to increase their market share not only by price competition (quite often set at a certain level by the cartel under the guidance of MITI in order to guarantee profitability), but also with stronger intensity in non-price competition, including the innovation of new products, efficient on-time delivery and various other services to meet the demands of customers.
4 The assistance and promotion of the iron and steel industry was central to government industrial policy in this period. As the success of political leadership was always determined by the evaluation of economic success, the major attention of government in the 1960s was on the achievement of heavy industrialisation rather than political status on the international scene. As was explained in Chapter 2, pp 24–6, the iron and steel industry benefited from various consistently favourable measures which enabled their large capital investment.
5 Large-scale iron and steel corporations became a key business unit in each financial group. The level of the Japanese iron and steel corporations' share capital was very low in 1968, at 20.4 per cent, but this handicap was compensated for by joining the financial group sharing interests with the major city banks. All financial groups centred around the major city banks made considerable efforts to create their financial hierarchy and the major iron and steel corporations were placed in the upper regions of each group. Thus much capital was made available for the modernisation of plant and equipment, and the channels for investment funds from particular banks to a particular steel corporation were also established, as examined in Chapter 5, pp 66–70.
6 The Japanese steel industry was successful in collecting the most advanced technologies from all over the world. As was explained in Chapter 5, pp. 74–9 and Figure 5.3 they collected them for each process from various countries, even including the USSR, and were able to operate the most efficient system of production.
7 Favourable labour–management relations based on the concept of enterprise-based industrial relations, which were established after 1950, encouraged management to invest a large amount of capital in modernisation to create green-field works at the most advantageous locations.

In contrast to the Japanese steel industry, the timing and amount of investment of the British steel industry was determined by the board of the Corporation, but had to be approved by the government of the time. Capital investment was concentrated in the 1970s, especially in the latter half, as was shown in Table 8.1; this represented a fifteen-year delay compared with that of Japan. The BSC's very high percentage (93.3 per cent at the peak period of 1974–6) in the total investment of the whole British steel industry was concentrated in the late 1970s. In the period prior to nationalisation in 1967 there had existed a reluctance to initiate large investment owing to the fear of nationalisation. When nationalisation happened, it took some years to reorganise the fourteen corporations into one integrated enterprise. When at last the *Ten Year Development Strategy* was drawn up, after the delay of the report by the Joint Steering Group, it was already 1973. There was also a delay in the execution of the above strategy due to the Beswick Review, which was a reflection of the industrial relations in the BSC. Three subsequent changes of government in 1970, 1974 and 1979 certainly imposed an inconsistent political influence and drag on the policies of the Corporation. Although the large-scale capital investment for the *Ten Year Development Strategy* was based on the demand forecast in the 1970s, as well as to cope with international competition, we might conclude that the late-1970s investment by the BSC was at a time when the great demand for steel was over and the general economic situation of the world was already one of recession.

The above difference in corporate environments was crucial for the evolution of two types of equipment modernisation: the Japanese type, a modernisation in expansion which reflects organic compatibility, and the British type, a modernisation in contraction which reflects contrived compatibility.

COMPARISON OF PLANT AND EQUIPMENT MODERNISATION

The Japanese modernisation was characterised by the establishment of many green-field works, while that of the UK was more or less partial and selective modernisation. The detail in scale and type of modernisation in Japan and the UK are in sharp contrast in terms of equipment in each process and the use of existing works. For example, as shown in Tables 8.1 and 11.1, there was a great increase in the number of blast furnaces in Japan, while in the UK they decreased greatly. However, with regard to blast furnace size, there has been a similar trend towards enlargement, which suggests that the larger-size furnace is better suited to the newly-introduced efficient steelmaking furnace (LD converter). In spite of some similar trends of modernisation, the difference in the growth of demand regulated the annual output of pig iron with great advantage to Japan. In 1965 there was no great difference in output between Japan and the UK (27.5 million tonnes for Japan and 17.7 million tonnes for the UK), but in 1980 the gap widened hugely (80.0 million tonnes for Japan and 6.4 million tonnes for the UK).

Table 11.1 Comparison of plant and equipment modernisation in the iron and steel industries, Japan and UK

Contents of equipment modernisation	*Japan*	*UK*
Capital investment		
period of intensive capital investment	1960s	1970s
Blast furnace		
Trend in quantitative change	Increase	Decrease
Number of furnaces (1980)	65	9
Trend in size	Enlargement	Enlargement
Maximum size	5,070m^3 Kashima Works (Sumitomo Metal Industries)	3,900m^3 Redcar (BSC)
LD converter		
Trend in quantitative change	Increase	Increase
Number of LD converters (1980)	94	16
Trend in size	Enlargement	Enlargement
Maximum size	340 tonnes Oita Works (Nippon Steel Corp.)	330 tonnes Port Talbot Works (BSC)
Continuous caster		
Trend in quantitative change	Increase	Increase
Number of casters (1980)	141	28
Hot strip mills		
Trend in quantitative change	Increase	Increase
Number of mills (1980)	23	6
Cold strip mills		
Trend in quantitative change	Increase	Slight increase
Number of mills (1980)	73	25
Automation		
Level	Extremely high	Medium
Areas	Comprehensive	Partial
Number of employees	Increase (1960–75) Decrease (1976–87)	Decrease (1960–87)

Sources: Data from Table 5.2 and Table 8.1; Tekkō Renmei, *Tōkei Yōran* 1981

As the benefits of LD converters became evident, the number and size of these converters increased in both countries to almost the same scale. The annual output of crude steel in 1965 was 41.1 million tonnes for Japan and 27.4 million tonnes for the UK; this is to be compared with that of 1980 – 111.3 million tonnes for Japan and 11.2 million tonnes for the UK. Such a huge gap in steel output despite similar modernisation is due to the difference in the size of demand. Continuous casters also increased in number in both countries, although the number of casters in Japan is greater than in the UK, reflecting the difference in the total volume of crude steel output.

A similar trend can be seen in the rolling process. The number of hot and cold strip mills have increased in both countries, though much less so in the UK. The use of computers has also advanced during the period 1960–80, but the way they are used varies between the two corporations. This difference in the use of automation at the corporation level is mainly due to the difference in the style of management, i.e. more centralisation in Japan and more decentralisation in the UK in terms of managerial control. Namely, in Japan a stronger and comprehensive concentration of control functions at head office has been achieved in the process of monolithic growth of organisations during the period of organic compatibility, while in the UK head office function is limited to the top level management functions and service functions common to all divisions. Managerial functions were largely transferred to the divisions, mainly due to the nationalisation of the fourteen different companies and the philosophy of top management at the stage of contrived compatibility.

Another important difference is that in Japan existing works were used fully, so as to recruit young and less union-conscious employees for the newly-established green-field works, which were placed at the most convenient sites for the market. In the UK all the modernisation and expansion programmes were carried out in the areas of existing works. Obviously there are a lot of advantages for green-field works in Japan, such as the Kimitsu works of Nippon Steel Corporation, and these are:

1 to select the best location for product markets as well as for receiving raw materials from overseas
2 to create the most efficient works layout
3 to assemble the most up-to-date equipment, supported by a high level of automation both at process and office levels
4 to choose and transfer young and non-union experienced employees from existing old works.

In the UK, for various reasons such as the small demand for steel products, a large number of existing relatively small works, and the social and political characteristics of steelworks regions which were reflections of traditional industrial relations institutions, it was difficult to establish a complete green-field works in the Japanese manner; thus the modernisation of the BSC could not but be more or less partial and selective.

Detailed comparison

A detailed comparison of the three large-scale works in Japan and the UK is given in Table 11.2. A comparison between the two modernised old plants (Yawata and Scunthorpe) and a green-field works (Kimitsu) will give us an indication of what similarities and differences can be recognised among them, and also how such differences affect management and labour. We can see that there are different ways of arranging plant and equipment to produce similar steel products from the same raw materials.

1 Location and layout of works: in comparing the three works, there is greater similarity in location and layout between Yawata and Scunthorpe than with Kimitsu. In other words, enlargement and modernisation of existing works (Yawata and Scunthorpe) tends to produce similarities, for they already have an established set of production factors, including workforces. By contrast, establishing a complete green-field works such as Kimitsu can produce unusually fortunate advantages under conditions of rapidly increasing demand.
2 Equipment and automation: we can find more similarity in the arrangement of equipment and level of automation between Yawata and Scunthorpe than Kimitsu works. The equipment installed at Kimitsu is large-scale, continuous, high-speed and automated, while in the Yawata and Scunthorpe works blast furnaces are smaller and automation is not all-on-line, but only has partial use of process computers. But one large difference between Yawata and Scunthorpe in modernisation is that Scunthorpe has three large LD converters (300 tonnes/charge), whereas Yawata has smaller LD converters, because they were installed at an earlier date.
3 Production and labour: the systems of production at the three works are all different. Kimitsu has a mass and multi-product production system as that site has a greater demand for a variety of products. The Yawata works has a smaller multi-product production system, as it already had a long history at a time when the demand was small

Table 11.2 Comparison of equipment modernisation in modern large-scale iron and steel works, Japan and UK

		Kimitsu works (1974)	*Yawata works (1974)*	*Scunthorpe works (1979)*
Location and layout	Location	Full use of advantages in coastal location directly connected with the largest market for demand	Enlargement and modernisation of the existing works, markets are distant (packed distribution)	Enlargement and modernisation of the existing works. Though it is said to be a coastal location, piers are situated 35.2 km away from the works
	Layout	Nationalisation of production processes: raw material–production–shipping, simplification of production line for each product	Irrational production process due to the combination of two separate sites	Layout is not rational particularly between iron and steel works
Equipment and automation	Equipment	Most modern large, continuous, high speed, automated equipment	Old, small capacity equipment, production lines are small in number	Selective modernisation: LD plant, continuous casters, bloom billet mill and medium section mill
	Blast furnace	Four furnaces 1 2,705 m^3 (1968) 2 1,540 m^3 (1969) 3 4,063 m^3 (1971) 4 4,930 m^3 (1975)	Six furnaces 1 1,020 m^3 (1930) 2 1,540 m^3 (1938) 3 1,913 m^3 (1959) 4 1,909 m^3 (1960) 5 2,338 m^3 (1962) 6 3,799 m^3 (1972)	Four furnaces 1 1,255 m^3 (1939) 2 1,385 m^3 (1939) 3 1,490 m^3 (1954) 4 1,510 m^3 (1954)

Table 11.2 Continued

		Kimitsu works (1974)	*Yawata works (1974)*	*Scunthorpe works (1979)*
	LD converter	Day capacity: 23,000 tonnes	Day capacity: 19,652 tonnes	Day capacity: 8,000 tonnes
		Five converters 1 3x250 tonnes/charge 2 2x300 tonnes/charge	Ten converters 1 1x50 tonnes/charge 2 2x60 tonnes/charge 3 2x75 tonnes/charge 4 5x150 tonnes/charge	Three converters 1 3x300 tonnes/charge
		Crude steel production capacity: 10 million tonnes/year	Crude steel production capacity: 8.9 million tonnes/year	Crude steel production capacity: 5.9 million tonnes/year
		Crude steel output: 6.76 million tonnes (1975)	Crude steel output: 6.88 million tonnes (1975)	Crude steel output: 3 million tonnes (1979) 2.87 million tonnes (1982)
	Automation	Comprehensive information system to control everything from order entry to shipping (39 process computers, 9 business computers)	No all-on-line system, process computers are partly used (26 process computers, 4 business computers)	No all-on-line system, partial use of process computers and business computers (8 process computers, 5 business computers)
Production and labour	Production system (products)	Mass and multi-products production system (plate, hot coil, cold coil, large sections, wire rod, galvanising zinc coating	Multi-products on small-scale production (rail, sheet-pile, wide flange beam, steel bar, plate, hot and cold rolled sheet,	Relatively limited products production system (plate, wire rod, medium and large sections). There is no hot

	coil, various tubes)	hot and cold coil, silicon steel sheets, tube, tin plate, galvanising zinc plate, tin-free steel, special steels, etc.)	strip mill, cold strip mill and no production of tubes
In-works sub-contracting system	Extensive in-works sub-contracting system (sub-contracting rate is 65.7 per cent in 1975)	Smaller scale sub-contracting system with a large number of sub-contracting firms (sub-contracting rate is 47.1 per cent in 1974)	Started in 1981 (sub-contracting rate is 7.4% in 1983)
Labour market	Supply from distant existing works by transfer of employees	Easy to secure labour in its own district	Difficult to obtain workforces from areas other than Scunthorpe
Number of employees	7,874 (April 1975) employees of sub-contracting firms: 13,904 (Dec. 1974)	21,575 (April 1975) employees of sub-contracting firms: 20,177 (Dec. 1974)	Decrease from 18,905 (Dec. 1979) to 9,815 (Dec. 1982)
Steel output per person	2.35 tonnes/day (1975)	0.87 tonnes/day (1975)	0.43 tonnes/day (1979) 0.87 tonnes/day (1982)

Sources: Scunthorpe works (BSC): Interview; Tekkō Renmei, *Tekkō no IE* 1975 (January); Kimitsu works (Nippon Steel): Interview

for each of various products. The Scunthorpe works has a production system with a relatively limited number of products, as it had already concentrated its efforts on particular products like plates and sections as a result of specialisation in products among major steel companies prior to nationalisation. In this sense there is already a decentralised corporate structure in the UK to which the Japanese steel companies may converge in future. In-works sub-contracting is an important element in the production system of both Kimitsu and Yawata works, with a higher percentage at Kimitsu. Advantages in terms of labour cost, flexibility in the use of workforce, benefits for stable industrial relations and other factors have already been considered in Chapter 5.

Sub-contracting in the Scunthorpe works was also initiated in 1980 and had increased to 7.3 per cent by 1983. The areas where sub-contracting was introduced in Scunthorpe are similar to Yawata and Kimitsu. They are bricklayers (47 people), refractory main hands (30), steel services (24), scrap preparation (10), welfare attendants (4) and office cleaner (one person) in 1981. In 1983 it was expanded further into the following areas: pyro-met (2 people), plate layers (107), joiners (29), painters (8), glaziers (6), salvage burners (7), torpedo repairs (45), welfare (125), canteens (107) and office cleaning (184) (Scunthorpe Works: Interview).

This sub-contracting suggests that similar conditions have now been created in the UK, which further suggests a trend of convergence of the BSC towards the Japanese steel companies. These conditions are that the benefits gained from the use of sub-contracting are similar to those of the Japanese companies; the trades unions have become more accommodated into the corporate strategy in order to create cost-efficient production systems; and there now seems to be availability of sub-contract workers in Britain, which, together with the second point, implies that there are now sub-contract companies in the UK ready to undertake various peripheral jobs in steelworks.

As to the labour market for the works, the conditions in Yawata and Scunthorpe are very similar in that they depend on the local workforce. Steel output per person was greatly increased in the Scunthorpe works from 1979 to 1982 due to the drastic decrease in the number of employees to the level of the Yawata works, although this does not indicate a comparative efficiency of the two LD plants, for various workers irrelevant to the steelmaking process are included in the calculation. Thus there is a greater similarity in the type of plant and equipment modernisation between the two existing large-scale works

(Yawata and Scunthorpe) than between either of those works and the green-field large works, Kimitsu.

COMPARISON OF MANAGEMENT

Top management

A major difference in top management between the Nippon Steel Corporation and the BSC is that in the former the decision-making and executive functions are in the hands of the same people, while in the latter they are separate.

The Nippon Steel Corporation as of 1978 has fifty-two directors spread over various rankings and divided into six hierarchical grades:

1 Representative director and chairman of the board of directors (one person);
2 Representative director and president (one person);
3 Director and executive vice president (10);
4 Managing directors (14);
5 Directors (22);
6 Standing auditors (2) and auditors (2).

All important decisions in both long- and mid-term policies, as well as major issues concerning current operations, are determined and executed by them. Some of the features of the top management of the Nippon Steel Corporation are:

1 The management policy council, consisting of one representative director and chairman of the board of directors, one representative director and president, and ten directors and executive vice presidents, plays the most important and substantial role in the top management function. Five vice presidents are responsible for multi-general functions which include: business planning, management of overseas activities, management of sub-contract affairs, general research, management of environment, affiliated activities, labour, general affairs, information systems, accounting, finance, sales, engineering business, raw material and fuel, machinery, and technology. Another five vice presidents are responsible for line functions, namely three modern large-scale works and one old large-scale works. Thus the group of vice presidents headed by the president and chairman controls the prime function of the top management.
2 Under this management policy council there is a group of managing

directors. Eight managing directors are responsible for general management functions and another six are in charge of line function.

3 Further under this group, there is a group of ordinary directors who are also divided into general managerial and line functions (see Table 7.4 p. 132).

Thus the top management of the Nippon Steel Corporation has three hierarchical structures which comprise the official board of directors. These directors are all those who have succeeded in climbing up the ladder of the huge and sophisticated company organisation through various jobs in sections and departments, finally to reach the top management class after forty to fifty years' service. This type of top management organisation is also a reflection of the people with the confidence and sense of responsibility to run the company for itself rather than for the benefit of shareholders – hence the need to establish good 'harmonious' social and industrial relations. These characteristics of top management remained almost the same even in 1994 when a new top management group was formed in response to the need for a drastic restructuring of corporate organisation. Cultural inheritance, reflected in the attitudes of the people, seems to lag behind the changing situation of the corporate environments.

In contrast to the Nippon Steel Corporation, the BSC's top management, although it has changed its structure a few times since nationalisation, has two different groups. These are the board, which is the official supreme decision-making organisation, and a group of executive managing directors who are delegated executive authority from the board. The board members, both full- and part-time, are appointed by the minister after consultation with the chairman of the Corporation, while the chairman himself is appointed by the minister. The number of full-time members has been declining since 1972, while that of part-time members increased remarkably in the period from 1972 to 1979 but showed a tendency to decrease towards 1982. It has been decided that some of the managing directors in charge of functions (such as finance and personnel) are qualified to become full-time board members, but not those in charge of line responsibility. Except for some full-time board members who are simultaneously managing directors, the other members are all appointed from outside the Corporation; in particular the chairman had always been externally inducted until in 1986 Mr R. Scholey was appointed. This structure of top management reflects the mode of ownership of the BSC, which was a public enterprise under the control of the government. However, basic BSC policies, affected by the changing political character of successive governments, have been inconsistent, which is in marked contrast to the experience of the Nippon Steel Corporation, which has benefited from the consistency deriving from its main policies being decided and executed by individuals who had risen to management level and internalised the aims of their predecessors. Thus the top management personnel in the Nippon Steel Corporation have had a stronger sense of being representative of and responsible for the fate of the whole corporation.

The second group of BSC's top management consisted of the managing directors, who are almost all promoted from within the Corporation. There has been almost no change in the number of functional managing directors, but that of the line increased slightly in 1979 in comparison with 1972 and in 1982. In 1982 two of the eight top management members responsible for the line were chairmen of the operating group, whose status ranked higher than that of the managing directors, and in addition one of the two chairmen was concurrently a full-time board member. Thus we can find some changes in the top management structure of the BSC which suggest a partial convergence towards the Japanese model in the sense that internally promoted people and those responsible for line functions gained more influence in top management functions in the 1980s. This characteristic has apparently been maintained even after the privatisation of BSC, as shown by the appointment of Brian Moffat, chairman and chief executive of British Steel plc. His service with the corporation was twenty-five years as of 1993; he became chief executive in 1991 and chairman in 1993.

Managerial hierarchy

At the Yawata and Scunthorpe works the total number of managerial employees above middle management was almost the same, the former having 140 and the latter 146 (see Tables 11.3 and 9.4, p. 192). In Yawata there were only two official positions, department manager and section chief, while in Scunthorpe there were three (as was shown in Table 9.4) – director, blue card and red card. Thus the managerial hierarchy has one more rank at Scunthorpe than Yawata. The difference in the level of hierarchy relates to the general character of the management organisation itself, i.e. in the flat hierarchy of management organisation at Yawata works, as seen in Table 11.4, specialisation of functions is more advanced at a higher level, namely at the departmental level with as many as twenty-five departments, while at Scunthorpe, this specialisation is very limited, with only seven at the corresponding level, which is a director position, as shown in Table 9.4. Why should this

Table 11.3 Composition of managerial employees by indirect and direct departments at Yawata and Kimitsu works (number and per cent)

	Yawata works (1960)			*Yawata works (1980)*			*Kimitsu works (1975)*		
Managerial position	DM	SC	Total	DM	SC	Total	DM	SC	Total
Indirect department	18	93	111 (86.0)	17	82	99 (67.8)	8	79	87 (71.9)
Direct department	3	15	18 (14.0)	8	39	47 (32.2)	6	28	34 (28.1)
Total	21 (16.3)	108 (83.7)	129 (100.0)	25 (17.1)	121 (82.9)	146 (100.0)	14 (11.6)	107 (88.4)	121 (100.0)

Source: Nippon Steel, 1980

Notes: (a) DM is a department manager and SC is a section chief

(b) Yawata works in 1960 was the main works of the Yawata Steel Corporation but the same works in 1980 was changed into one of the major works (establishment) of the Nippon Steel Corporation due to the merger of the Yawata and the Fuji Corporations in 1970

(c) In both periods at Yawata works and in 1975 at Kimitsu works there were one head and two deputy heads of the works and they were all members of the board of the Corporation but they are not included in this figure

Table 11.4 Composition of managerial employees by direct and indirect departments (Yawata Works, 1980)

Organisation	*Department manager*	*Section chief*	*Total*
Indirect department			
Personnel	1	1	2
General affairs	1	2	3
System development	1	3	4
Environmental control	1	2	3
Labour	1	5	6
Education, training	1	1	2
Accounting	1	4	5
Purchase	1	4	5
Technical	1	10	11
Experimentation	1	3	4
Technical laboratory	1	0	1
Production operations	1	4	5
Distribution management	1	2	3
Energy department	1	4	5
Equipment department	1	17	18
Technical assistance	1	0	1
Hospital	1	20	21
Sub-total	17	82	99 (67.8)
Direct department			
Iron making	1	7	8
Steel making	1	6	7
Plate and bar steel	1	5	6
Sheet	1	6	7
Surface treatment	1	3	4
Magnetic plate	1	3	4
Seamless tube	1	6	7
Nakatsu factory	1	3	4
Sub-total	8	39	47 (32.2)
Total	25 (17.1)	121 (82.9)	146 (100.0)

Source: Nippon Steel Corporation 1980
Note: There is one head of works whose position is a vice president and two deputy heads whose positions are directors on the board of the Corporation

difference in the structure of management exist while the total number of managers is almost the same? Logically we can say that it is because of the greater specialisation and division of labour in the Yawata works at the higher managerial level. There, functions such as general affairs, system development, environmental control, labour, education and training, distribution management, energy department, equipment department, and hospital are all regarded as having the same importance as those basic functions which exist in the Scunthorpe works, such as finance control, personnel and social policy, engineering, supplies and production control and so on. In the direct departments as well, more specialisation is made and in consequence a higher status is given to each line function of the direct departments. In the Yawata works ironmaking, steelmaking, plate and bar steel and similar line functions are all given a higher status and are treated as equal.

Such difference in management organisation is partly due to the difference in the concept of forming organisations, which can be traced back to the difference in the style of employment contract. This can be summarised as a flexible career-based contract in Japan, as against a rigid job-based contract in Britain. The Japanese system is based upon a collection of various organisations consisting of departments and sections in which individuals are regarded as a part of the collective whole, rather than flexibly given a job, and then share the function of the organisation. This system makes rules regarding delegation of authority and reporting to a superior much looser and less individualistic than in the BSC. Instead, regular meetings of departmental managers take place in order to exchange ideas and opinions, which help to foster a participative climate among managers at this level. This aspect of social structure reflected in management organisation makes it possible for the head and two deputy heads of the works to control and represent a large number of departmental managers, who can delegate their substantial responsibility to the section chiefs. By contrast, in the BSC the management organisation is composed of and built upon a steep mass of separate jobs to which each individual is attached. Thus there is a clear demarcation in both lateral and vertical functions, grouped as jobs, as seen on the charts of individuals, linked and piled up in the whole management organisation. It is difficult to make a flat organisation under this organisational paradigm; rather, there always exists a tendency for the official hierarchy to develop higher in the management organisation. This difference in the concept of management organisation does not mean that managerial employees in the Japanese corporation are not graded hierarchically, but that the employees in the Nippon Steel Corporation management are more finely ranked, related to slight differences in salary and seniority, which also creates a climate of status consciousness among employees. Although the current characteristics of managerial organisation are not directly related to the institutional arrangements of the pre-war years, the Nippon Steel Corporation owes its original managerial concept to the introduction of the Japanese civil service organisation (1901) (Nippon Steel 1980: 36), and the influence of *Shokumukyu Seido* recommended by the US to the Japanese civil service organisation (1946).[1]

Direct and indirect department

In the ratio of managerial employees between direct and indirect departments in 1980/81 there is not much difference between Yawata and Scunthorpe (see Tables 11.3 and 9.4). When compared with the green-field works at Kimitsu, we see that the ratio of the Kimitsu works is closer to Scunthorpe than to Yawata. This suggests how the large-scale re-organisation was carried out in the Scunthorpe works and numerous staff functions included in the direct (line) department in 1974 were transferred to the indirect (function) department up to the Kimitsu level. Table 9.4 (see p. 192) shows that the management organisation of the Scunthorpe works in 1974 had a ratio of indirect and direct departments completely opposite to that of 1981. This indicates that a similar type of equipment modernisation forces management to introduce a basically similar personnel re-organisation of the works, although there still exist some minor differences.

Labour management (Japan)

The key strategy for labour management in the Nippon Steel Corporation has been to develop education and training and use them for two purposes. One is to use the system to provide employees with the knowledge and skills required for the job they are assigned to, and the other is to use it for assessment of the employees. With these two methods they aimed to achieve the most efficient allocation and use of the workforce in the enterprise. Many old jobs based upon traditional skills have been replaced by more standardised and knowledge-based jobs in the main processes and this shift has made the in-firm education and training significant as well as useful, especially when organisational growth is in progress, as it can play an important role as a strategy of management. In the life-time employment and seniority systems there always arise discrepancies between managerial positions and employees of the same seniority and it is crucial to adjust these and have a fair and efficient method of solving this problem. Assessment through the education and training processes has been used as a means for solving this problem, although it has not been as fair as it should be. In practice, objective results of in-firm education and training have been modified by the subjective consideration of managers according to their personal connections, academic clique, ideological views, and so on. In general, the system was basically applied to blue-collar employees and up to *Sanji* qualification for white-collar employees. Above *Sanji*, evaluation becomes more personal and subjective, and thus the educational and training schema (Figure 7.2, p. 149) as a means of personnel management becomes more or less an official plan made up by the *Nōryoku Kaihatsubu* (faculty development department).

The following are the features identified in the Japanese in-firm education and training programmes, which also contributed to 'harmonious' industrial relations during the high-growth period:

1 Introductory education and training is given according to previous academic qualifications, namely according to the level and type of education received in schools and universities. Thus employees are roughly classified at the start of their life-time employment as to which path in the organisation they will take.
2 The period and training for ordinary employees has become shorter in general due to the effects of plant and equipment modernisation as well as the various sources of selective recruitment, such as graduates of postgraduate courses, universities, iron and steel colleges, technical colleges and women's colleges, as well as male and female high school graduates. The higher the school career, the longer the programme; the head office has a shorter programme than the steel works, the new steel works has a shorter programme than the old steel works, female high school graduates have a shorter programme than male high school graduates.
3 The type of education is hierarchically structured and linked with promotion both in qualification ranks and line posts. Education and training programmes are planned in the long-term and their cost is regarded as an investment in increasing the value of the workforce.

The education and training programmes provide employees with a general awareness of the company and its role in the industry and the country, as well as the knowledge and skills they need in their immediate jobs, thus strengthening the idea that they are members of the company rather than simply employed for a particular job. There is always an assumption that an employee should be regarded as a person who co-operates in the team and works towards the success of the organisation, thus fostering the pride of being an individual working for collective goals. Moral aspects, such as the importance of self-improvement and loyalty to one's superior, are always emphasised with the implicit philosophy of Confucianism as a prerequisite for successful labour management. Leaders are expected to become rich in personality and lead and co-ordinate their subordinates rather than showing strong leadership with apparent power and ability to control. Among the abilities required of managerial personnel, professional managerial knowledge and techniques, human relations and good leadership are considered to be more important than other elements.

The importance of *Jishukanri Katsudō* (self-management activity) and the safety programme increased as an important managerial strategy which was a part of labour management in the 1970s, when the basic plant and equipment modernisation was completed, in order to optimise team incentives. This has become both more possible and necessary as a result of increasing the de-skilling tendencies under automation and also due to the necessity of creating various minor improvements to increase efficiency and rationalisation in the workplace.

The education and training of a foreman remains most important; he is regarded as a key figure in labour management, acting as a co-ordinator between workers and management. Particular emphasis has been put on the education and training of section chiefs, which have increased both in number and percentage. The study and training for section chiefs include scientific management, scientific thinking in decision-making, professional knowledge and higher-level skills to solve various managerial problems, and to give them a broad vision and knowledge about business management through various meetings with managers and directors, as well as attending seminars held outside the company.

Employee education is linked to the *Shokunō Shikaku Seido*, which took on its complete form in the early 1970s. This is an important change which can be regarded as evidence of convergence, Japan moving in the direction of individualistic Western employment practices, as will be explained further in the next chapter. The *Shokunō Shikaku Seido*, as shown in Figure 7.2, and its detailed method of usage is a means of allocating and adjusting the number of employees to various qualification ranks and line posts; it is thus the most important tool in maintaining efficient control over employees.

The implications of the above characteristics of education and training in the Nippon Steel Corporation in contrast to that of the BSC are:

1 In-firm education and training became necessary because the development of the industry after 1955 was so fast and large in scale and

moreover continued for a period of twenty years. This unique corporate environment, which was a combination of two periods – one a period of Japanese heavy industrialisation and the other a world-wide economic boom – made it difficult to recruit already-existing skilled steel workers and it reflects a feature of the late and rapid industrialisation of Japan. It resembles the first systematic in-firm education and training which was initiated as early as 1910 at Yawata works, then under government ownership. On the contrary, in the BSC the necessity of systematic in-firm education and training was not so strongly required because:

(a) there existed a sufficient workforce already trained outside the enterprise except for some kinds of craftsmen which were from time to time in short supply;
(b) as employees, particularly manual workers, are employed by contract in relation to a particular job with a rather high turnover rate, there was no necessity to educate and train them as employees who would spend their whole working life in the same enterprise;
(c) although the modernisation of plant and equipment required employees with new knowledge and skills, it was not necessary to establish a systematic in-firm education and training as it indirectly hinders any strategy of drastic reduction of the workforce, which had been an overriding issue at BSC since nationalisation in 1967.

2 As top management in Japan had a long-term view concerning the growth of the enterprise and as employees had a similar view of their future in the company, there was no sense of seeing money spent on training and education as a cost; it was, rather, regarded as an investment in human capital for the future development of the company. But in the BSC it has generally been seen as an expense, as reflected in the far less intensive and non-standardised off-the-job training, except for some relatively systematic education for managerial employees, who are expected to become core managerial employees. Even so, it is not used as part of a long-term career system; university graduates, for example, are expected to undertake a managerial post after one or two years, while in the Nippon Steel Corporation their first ten years are mostly spent as ordinary staff members before becoming future managers of various functions.

3 In-firm training and education has become a useful means of evaluating employees and allocating them to various functions or jobs and also to qualification ranks and line posts as an important measure of personnel management.

The whole system of training and education in the enterprise has been matched to, and been made an extension of, general Japanese social values characteristic of a rising industrial society, and regarded as a useful and reasonable method of evaluating individuals. At the same time, combined with the effect of the seniority practice and promotion system in the enterprise, it contributed greatly to making the values and attitudes of employees suit the idea of enterprise-based 'harmonious' industrial relations.

Labour management (UK)

The basic labour management practice of the BSC is based on a traditional method of classifying and separating similar functions into jobs and occupations and then evaluating them in terms of importance and scarcity of the workforce and linking them to levels of wage/salary. This basic notion of job–wage relationship dates as far back as the beginning of the nineteenth century when the number of factories increased during the process of industrialisation. Even now, there is a remaining influence as seen in the clear distinction between blue-collar workers, who are paid weekly, and white-collar staff and management, who are paid on a monthly basis, though this is likely to change in the future to an integrated payment system.

The summary in Table 11.5 of the relationship between changes in the number of blue-collar employees and the level of their wages/salaries, reflects the combined effects of technology and the strategy of labour management within the framework of traditional management in the BSC. The following three points can be adduced from Table 11.5:

1. Occupations and jobs which decreased largely as a result of plant and equipment modernisation (the content of such jobs are standardised, simplified and now changed to a typical automated type of job) have a low wage/salary level and their incrementation is intermediate or low.
2. Wage/salary level and incrementation are either intermediate or high for occupations and jobs which did not decrease by much, but rather increased in importance due to plant and equipment modernisation.
3. Newly-created jobs, due to the modernisation of plant and equipment, such as in a continuous caster plant, have a high level of reward and the rate of increase is also high.

Thus we can confirm that the long-term trend in the differentiation of wage/salary level in the BSC is relevant to the change in the content of jobs created by technological changes, which will eventually affect the demand and supply conditions for particular jobs, although the wage/salary rate at a particular time is settled by negotiations between

Table 11.5 Changes in the number of white-collar and blue-collar employees in major occupations and their relations to changes in wages/salaries (BSC, 1972–9)

Employees	*Decrease or increase in employment (1972–9)%*	*Level of wage/salary%*	*Increase rate of wage/salary (1972–9)%*
White-collar (male):		Male clerical supervisor (100.0)	
Clerical	Large decrease (−33.8)	Low (97.1)	Intermediate (188.2)
Technicians (mechanical)	Small decrease (−9.7)	High (107.2)	High (197.6)
White-collar (female):		Female clerical supervisor (100.0)	
Clerical	Increase (+23.6)	Intermediate (75.7)	Intermediate (212.2)
Typists, etc.	Large decrease (−35.6)	Low (75.2)	Low (208.7)
Process operatives:		Male foreman (100.0)	
Cold reduction mill	Large decrease (−43.2)	Low (70.7)	Low (169.4)
Ancillary departments (coating, etc.)	Small decrease (−5.9)	High (75.8)	High (185.1)
Continuous casting plants	Increase (+85.2)	High (84.8)	High (215.8)
Skilled employees:		Male foreman (100.0)	
Blacksmith	Large decrease (−42.2)	Low (79.7)	Intermediate (177.0)
Electricians, fitters	Small decrease (−14.2)	High (87.3)	High (179.4)
Ancillary employees:		Male foreman (100.0)	
General labourers, rail drivers and shunters	Large decrease (−41.1)	Intermediate (65.6)	Intermediate (208.2)
Crane drivers and slingers, road drivers	Small decrease (−5.5)	High (78.2)	Intermediate (179.3)

Source: Compiled from Tables 8.4 to 8.13 in Chapter 8
Note: Adjectives such as large, small, high, low and intermediate are given as a result of relative comparisons of percentages among all occupations in each category. Detailed data for all occupations are given in Tables 8.4 to 8.13

management and trade unions which represent employees of certain jobs and occupations. Such basic technological and economic factors affecting long-term trends in the differentiation of the level of salaries will also be observed in the Japanese corporations if the level of salaries by jobs and occupations is analysed. But in Japan factors such as general educational career and individual in-firm promotion systems take precedence over the effect of horizontal industrially-based negotiations as the employees in the major steel corporations are not employed in relation to particular jobs and occupations, as shown in Tables 11.6 and 11.7.

Table 11.6 Differences in salary level in different managerial positions (BSC, 1983)

Managerial positions	*Differentials %*	
Staff	100.0	
Stoveminder (blue-collar)	101.3	
Foreman	102.9	(Grades 1 to 4 depending upon specific foreman position, and increase in length of service up to four years)
Keeper (blue-collar)	113.2	
Middle management	168.5	(Grades from MM-0 to MM-6)
Management	222.7	(Grades from M-0 to M-4)
Senior management	390.2	
Top management	650.4	

Sources: BSC Head Office: Interview; Scunthorpe Works (BSC): Interview

Table 11.7 Differences in salary level in different managerial positions (Nippon Steel Corporation, 1978)

Average age	*Managerial positions*	*Differentials %*
22–23	Ordinary staff	100.0
40	Assistant section chief	263.2
45	Section chief	299.5
50	Assistant to department manager	323.7
55	Department manager	376.2
	(Non-department manager)	353.1
Above 60	Top management	846.1

Sources: Ishida 1981: 268; Nippon Steel, *Yukashōken Hōkokusho* 1978
Note: Figures are calculated based on the monthly rate of the head office managerial salaries of the five major iron and steel corporations. The data for top management salaries is from the Nippon Steel Corporation, announced in *Yukashōken Hōkokusho* (financial statement)

The following points can be noted as characteristics of the management of white-collar employees and managers at the BSC, as compared to that of the Nippon Steel Corporation.

First, the initial level of salary in the BSC is in general large enough to support a family (this applies to blue-collar workers as well), while in Japan it is set at a level only high enough to support a single person, and the level of salary goes up later as the worker gets older and his family expenditure gradually increases (Table 11.8). The basic level of salary in the BSC means that the salary of ordinary staff is set relatively low and with relatively little differential, compared with managerial employees, for whom the level of salary is personal, and 'merit' is stronger as a determinant of salaries. Thus the system is one of less incentive for ordinary staff, but higher incentive for managerial employees.

Second, in the BSC the level of salary differs depending upon job, occupation, and managerial position. In each managerial position there are grades, such as four grades for foreman, seven grades for middle management and five grades for management (Table 11.6). In addition there are several ranks in each grade, and staff and managerial employees are automatically promoted to a higher rank in each grade depending on their length of service, thus providing the conditions for establishing a more flexible job structure and an incentive for a longer working life in the same corporation.

Third, in terms of relative pay differentials, senior management in the BSC seems to be better paid than its counterpart in the Nippon Steel

Table 11.8 Wage/salaries in a selected major steel company, Japan

University graduates, white-collar (1978)		*High school graduates, blue-collar (1978)*	
Age	*%*	*Age*	*%*
22	100.0	22	100.0
25	123.5	25	116.3
30	165.6	30	136.3
35	212.4	35	160.1
40	247.8	40	180.8
45	289.9	45	195.9
50	322.5	50	210.6
55	364.3	55	228.6

Source: Kato 1981

Note: The amount of salary is the average model salary which includes the basic rate, incentive rate and other allowances

Corporation (department manager), while top management in the Nippon Steel Corporation is more highly paid than its counterpart in the BSC. But the relatively high level of salary of Japanese top management is due to comparison with the salary they received at the age of 22–23, at a level only high enough to support a single person. In Japan the salary basically goes up as one gets older, and the differential increases by seniority rather than the position held, as is shown in the level of salary for a 55-year-old department manager and non-department manager of the same age (Table 11.6).

The above differences described as characteristics of the wage/salary systems in Japan and the UK derive from variance in the basic principle of employment itself, namely the notion of life-time employment in Japan and job-contractual employment in the UK. The principles found in our research cannot be understood without reference to the process of labour market formation in the two countries; the early and gradual development of capitalism in the UK and the late and rapid development in Japan. This does not imply that the principles have no rationale under current conditions of the respective labour markets, although they may not be at the most compatible situation as regards management. Efforts towards modification are being made in both cases in order to maintain management compatibility.

In the Japanese case there are many complicated methods to modify the simple seniority system, but the general amount of salary shown in Table 11.8 indicates that a finely graded salary scale still remains based upon seniority (age, length of service) in the revised life-time employment practice under *Shokunō Shikaku Seido*. It is difficult to make a normative judgement, but the Japanese salary structure, in which the difference in salary is finely graded and the salary differential is in general smaller and based upon seniority throughout the corporate system, seems to be more convenient for obtaining a stronger collective force for the organisational objective. In this sense it is a phenomenon of management at a stage of organic compatibility in Japan. The Japanese structure gives the employees a combined sense of equality, security and competition among themselves. A sense of competition has been created by a slight differentiation in the evaluation of individuals by their superiors throughout the whole in-firm career system. It worked very well in the high economic growth period of the 1960s and was maintained with some modifications at least up to the 1980s in Japan. However, in the BSC the old institutions, established under the corporate environment in the nineteenth century and modified since then, now stand as rigid obstacles against institutional innovation. In spite of that, in the 1980s the effects of the plant and equipment modernisation achieved in the 1970s have produced favourable conditions to the point where management can negotiate with the trade unions in order to change traditional institutions towards a situation more favourable for management.

The role of the foreman

There are distinct differences in the role of foreman between the major Japanese steel companies and the BSC, as summarised below. These are all interrelated and relevant to the difference in the degree of managerial importance invested in the post. A considerably higher status accorded to Japanese foremen, who act as core workplace management personnel, has been a source of good industrial relations in Japan, while the fact that the status of foreman of the BSC was under review in 1987 suggests a change in the approach to workplace industrial relations in the UK.

The average age of foremen in the major iron and steel corporations in Japan is relatively high at 42, while in the BSC it is 30–35 (Table 11.9). The average length of service in Japan is relatively long, amounting to twenty-four years, while in the BSC it is relatively short, the minimum requirement being five years (Table 11.9). The salary of a foreman in Japan is relatively high, in fact 34.5 per cent higher than that of ordinary workers of the same age, while in the BSC the wage of the keeper is higher than that of the foreman (Table 11.9). As was examined in Chapter 6, pp. 121–7 one can become a foreman in Japan after long in-firm education and training, and interviews. The works manager recommends a group leader with a good character, excellent skills, good work performance, plus strong leadership with a full understanding of the feelings and psychology of the workers. Thus the method of appointment reflects the type of foreman required for successful labour management at the workplace. However, in the BSC the job of foreman is considered to be just one of the many jobs in the Corporation, as shown in the method of appointment, so it is relatively easy to attain the position. There are two methods of appointment: vacancies for established foreman positions are normally advertised in the works newspaper, unless they arise as a result of a re-organisation either across the works or within a department, in which case selection is made from existing foremen, usually with some union involvement. A second mode of appointment at Scunthorpe has traditionally derived from the supervisory training scheme. In this case applications are invited, in the works newspaper, for places on the training scheme. In Japan recruitment is only from among the blue-collar workers, while in the BSC, graduate trainees are generally encouraged to apply for vacancies as part of their development, especially in production areas. The education and training in Japan is continuous and systematic even before and after becoming a foreman, while in the BSC education and training for foremen is given only after appointment; thus it is relatively short and not systematic. The Japanese foreman is given a more important role and authority in labour management in addition to work management, while in the BSC the job is mostly limited to work management, with less responsibility for the labour management function, especially in human relations management in the workplace. In Japan this is relatively large, being nineteen workers on average (1980), while in the BSC it is fifteen workers (1979). In Japan, however, labour management at the end of the hierarchy is enforced by the use of group leaders (Table 11.9). The relationship with workers in Japan is an harmonious paternal one, as a foreman is immersed in the work team and is a respected boss of the team. Seniority (age and length of service) is maintained in Japan in order to create and keep good human relationships, which conforms to the general values of a society in which seniority is respected, and traditionally based upon Confucian attitudes and values. But in the BSC it is a conflictual and power-conscious relationship, as a foreman's legitimacy has become one based upon power rather than authority. This is due to two factors: first, the foreman is not considered a member of the work team; second, workers have their shop steward, who is also a worker, and can negotiate with management at levels considerably senior to first-line supervisors and foremen. In the Nippon Steel Corporation the foremen are not members of the union, while in the BSC most of them are members either of the ISTC or SIMA.

The above differences suggest that the Japanese foreman does have a more important role in labour management and his status as the first line of management is considerably higher than that of his British counterpart. Marsh (1979: 117) points to four factors leading to the decline in status of the foreman in the UK. They are the growth of mass production; trade union development and the role of the shop steward; the proliferation of staff specialists arising out of expanding technology and scientific management; and the hierarchical structures of modern management. As the major steel companies in Japan share all these factors except the second one, we might assume that the second factor is most relevant to the difference in the role and status of a foreman between the major steel companies in Japan and the BSC. In Japan, the process of late and rapid industrialisation, and a strong management initiative supported by the ideology of economic nationalism have helped to make trade union attitudes co-operative. Under such circumstances, foremen have been given greater authority, including that of labour management, and especially the right to evaluate promotion; in addition foremen were not allowed to become members of a union. Such a historical difference has made foremen more authoritative and they have succeeded in directing their new workers with no trade union experience towards the objective of the enterprise. On the other hand, in the BSC the managerial importance of foremen has declined for the following four reasons: promotion prospects are overwhelmingly determined by seniority, as specified in the union rule-book, rather than by unilateral decision of the management; the union representative (shop-steward) often effectively has a higher status than the foremen, reflecting the fact that he has to negotiate in collective bargaining with management levels considerably higher than foremen; the work management duties of foremen have also been split partly downward to the senior blue-collar jobholders and partly upward to the lower ranks of middle management; and finally, the foremen have also been unionised in order to protect their own positions, which makes them regard the maintenance of their power vis-à-vis the workers as the main focus of their attention, so that organisational goals may become secondary as far as the foreman is concerned.

Thus, managerial authority and control at the workplace has been maintained in the hands of foremen in the Japanese steel companies, making it easier for the management to increase managerial authority at the workplace level. In contrast, the role of foremen in the BSC has been viewed quite differently, and this will be considered in the next chapter as an aspect of changing trends.

COMPARISON OF LABOUR

The introduction of common technology tends to create a common change in the size and composition of a workforce under the twin pressures of competition and efficiency. Although in details they differ, in the general and long-term trend there seems to be a convergence in the use of the workforce both quantitatively and qualitatively. The type of labour predominating may then affect the structure and functions of industrial relations institutions, as we will see in the final chapter.

The workforce at the workplace

We can assume that similar technology encourages the use of a similar number of workers. But in reality it is nearly impossible to find exactly the same technology used in the same technical and managerial manner, and thus there are varying numbers of workers organised in different ways. Yet in spite of such small-scale disparity a basic similarity may still be identified in the use of equipment and the size of the workforce. The workforce composition at the workplace in both the Japanese and the British cases was examined separately in Chapters 6 and 9. A more detailed comparison will be made here to examine specific differences and similarities using information from the Kure works of Nisshin Steel Co. Ltd, since the similar-sized blast furnace in the Yawata works was not in use at the time of this research. An outline of the two works follows.

Blast furnace No. 2 at Kure was built in 1966 with an annual capacity of 1.22 million tonnes and an inner volume of 1,650 m^3, while Furnace No. 4 at Scunthorpe was built in 1954 with an annual capacity of 0.97 million tonnes and an inner volume of 1,510 m^3. The workforce composition already given in Table 9.6 (p. 197) has here been integrated into Table 11.9 for comparison with details of a similar team at Kure works. This gives us some useful knowledge about the relationship between the equipment, workforce composition and labour management.

1 The number of workers: there are nine workers, including the foreman, at the Scunthorpe works, while there are twelve at Kure. The difference in number can be ascribed to two factors. One is that Kure has more workers in the hob work group, because the blast furnace has two hot metal drainages. The other is that Kure has more people in the hot stove and crane work team, because they are in charge of two furnaces in the same control room (Nos. 1 and 2 furnaces are adjacent). Hence, even with almost identical equipment, they need a different number of workers owing to the specific differences in the usage of the equipment itself and its arrangement in relation to other equipment. In other words it conversely supports the idea that a similar number of workers in the workplace reflects the use of similar technology. It also suggests that the same technology requires the same number of workers to operate it if there are no social restraints such as an accepted standard of working conditions, which is determined by the forces of industrial relations regarding the use of workers, or any unique managerial philosophy peculiar to that enterprise.
2 Promotion of workers: a difference in the arrangement of workers between the two workplaces is that the hob work team in the Scunthorpe works has a more hierarchical arrangement, with the titles of keeper, first, second, third and fourth helpers, while the Kure

Table 11.9 Comparison of the workforce composition of blast furnace teams (Kure works, 1983 and Scunthorpe works, 1984)

	Average age	*Average length of service (years)*	*School career*	*Salary (Japan ¥) wage (UK £/week)*
Japan: blast furnace hob work				
Foreman	42	24	High School	373,611
Group leader	46	24	High School	323,611
Senior grade	43	18	High School	278,011
Ordinary workers A	38	14	High School	253,826
B				
C				
D				
Japan: blast furnace control room				
Group leader	44	22	High School	319,859
Senior grade	39	21	High School	309,194
Ordinary workers A	30	10	High School	219,341
B				
C				
UK: blast furnace hob work				
Foreman	30–35	5 (min.)	Various	152.00
Keeper	45–55 (52)	10 (min.) (24)	Secondary	167.38
1st helper	18–55 (45)	0–10 (17)	Secondary	149.75
2nd helper	18–55 (38)	0–10 (9)	Secondary	149.75
3rd helper	18–55 (38)	0–10 (9)	Secondary	143.34
4th helper	18–55 (34)	0–10 (7)	Secondary	143.34
UK: blast furnace other work				
Stoveminder	18–55 (38)	0–10 (9)	Secondary	149.75
Crane driver	18–55 (34)	0–10 (7)	Secondary	132.14
Charging attendent	18–55 (38)	0–10 (9)	Secondary	154.46

Sources: Kure works (Nisshin Steel Co., Ltd): Interview; Scunthorpe works (BSC): Interview

Notes: (a) The high school finishing age in Japan is 18 and secondary school in the UK is minimum 16. Secondary school also includes sixth form, 17–18 years old

(b) ( ) in the Scunthorpe works shows the actual figures for 1981

works has four workers who each hold the title of 'ordinary' worker, which does not specify any official rank or job, and this ensures flexibility in terms of responsibility for the tasks in the workplace. In the area of hot stove and crane work the workers in the Scunthorpe works do have their own specific job titles. In contrast to a rather rigid job-centred work organisation and apparent hierarchy at Scunthorpe, the Kure situation is looser and more flexible in the use of workers, as these four workers are grouped under a common job title. In addition a group leader and a senior grade worker are willing to help ordinary workers without any sense of job demarcation. Their willingness to help them is also counted as an important human relations factor required of foremen, thus making for successful team performance.

3 Age and length of service: the salient difference and implication observed between the two workplaces is that at Scunthorpe the length of service, except for that of the keeper, is relatively short while ages are quite high, thus suggesting that all the workers, including the keeper, entered this workplace in the middle of their working career; in other words, it reflects a higher turnover rate in general industry. In contrast, in Japan most of the workers took up their employment as soon as, or within one or two years after, they finished high school education. It gives us the impression that in the Kure works length of service is considered more important than age itself if other conditions for promotion are almost equal (as is seen in the reverse age seniority between foreman and group leader).

4 Educational level and wage/salary: the major difference between Japan and the UK is that the workers in the UK example all left secondary school at the age of 16, while the Japanese workers graduated from high school at 18. In Britain primary education begins at the age of 5 and in Japan at the age of 6. The difference in the length of education is therefore only one year but schooling from 16–18 is, from the point of view of future career, more important than 5–6 in terms of the amount and kinds of knowledge received. In addition, a comparison of the content of secondary education in the UK and that of Japanese high schools is perhaps important. Japanese high school education is more uniform in curriculum and centred on knowledge of various subjects required for the entrance examinations to colleges and universities. This type of wide-ranging knowledge-intensive education perhaps better prepares the would-be worker to follow the in-enterprise systematic education and training examined in Chapter 6, pp. 121–7 when the production system itself is already automated. The in-enterprise education and training programmes

which aim at providing workers with more comprehensive and greater knowledge of technical systems of steelmaking and the whole technical system of the works will be rather more easily accepted; it will also help workers to understand the importance and role of their present job and to be more flexible in moving from one job to another with the aim of broadening experience within the company.

As to wages and salary, the difference between the British and Japanese systems is obvious in two ways: first British workers are paid weekly, while Japanese workers are paid on a monthly basis in all staff categories, thus having a stronger sense of equality in the method of payment, which leads to an element of good industrial relations among employees. Second, the principle of payment is apparently different in that British workers are paid for the job, almost regardless of age and length of service. For example, as is shown in Table 11.9, the first helper in a hob work team, a man of 45 years of age with seventeen years' service, received the same wage as the second helper who was seven years younger with eight years' less service. This would never happen in a Japanese hob work team. In addition the difference in wage between different jobs in the UK is only a matter of a few pounds.

In a Japanese steelworks, as shown in Table 11.9, age and length of service correspond closely and the salary scale is based upon the seniority principle (age and length of service), with a much larger differential between each salary than in the UK. This means that the initial salary of young workers in Japan is set at a level to support life as a single householder/person and then the curve of salary scale rises so as to reach a peak when family expenditure is likely to maximise. In Japan, therefore, it may happen that a foreman who is younger and with a shorter length of service receives a lower salary than an ordinary, older worker with more service. This is not usually the case in the same workplace, as it damages human relationships among workers. But in the newly-established works such as Kimitsu, younger foremen with relatively short service have a lower salary than that of ordinary older workers in the Yawata works.

The significance of such Japanese seniority practices in relation to labour management and industrial relations are, first, that about half of the total salary is determined by a finely graded differentiation made by seniority ranking; this provides employees with a sense of assurance and satisfaction which accrues from the egalitarian principle of this practice. They feel as if they are going up a set of stairs step by step with forty or forty-five steps depending upon the number of years to their retirement. As their seniority position goes up annually, all employees are brought to recognise their own seniority status among employees, according to their ranking. The sense of assurance and satisfaction helps to create a climate of corporate community and enhances their reliability for the enterprise, and thereby a stronger sense of co-operation among employees and participation in management objectives as long as the growth of the company continues.

Second, from a more practical aspect, this practice contributes to the formation of good industrial relations, for once an employee joins the enterprise and goes up to a medium ranking of seniority, at 35–40 years of age, there is a clear disadvantage in changing company, because the new employer would place them at a lower point in its own seniority table in order not to disturb their own length of service status quo. This, together with the first consideration, helps to create a psychological element to enterprise-based industrial relations.

The workforce at the works

The similar size and composition of the workforce and their changing trends at the works is also a manifestation of technology, as evidenced in an increased similarity between the Yawata and Scunthorpe works, rather than the Kimitsu works, but the detailed differences in the two old works are considered due to factors other than technology, although the degree of influence of such factors is difficult to quantify.

Nippon Steel Corporation has nine steel works, with Yawata being the largest and Kimitsu one of the most efficient green-field works. In contrast, the British Steel Corporation has six large-scale iron and steel works and Scunthorpe is one of the most modernised large-scale sites. As it is almost impossible to find exactly identical cases for comparison at the establishment level in both Japan and the UK, the two most appropriate works in terms of general technological conditions were chosen. A comparison of workforce composition between the Yawata works in 1980 and the Scunthorpe works in 1981 was made in order to see how they differed after modernisation; they are then compared with the most modern Japanese green-field works – Kimitsu (see Table 11.10). At the same time, the workforce composition of Yawata in 1980 and Scunthorpe in 1981 are compared with those of 1971 and 1972 respectively in order to elucidate the impact of modernisation (see Table 11.11).

The percentage of managerial employees at the Scunthorpe works is larger than at either Yawata or Kimitsu. This suggests that in the BSC the degree of decentralisation is more advanced and in consequence

Table 11.10 Comparison of workforce composition of representative large-scale iron and steel works (Japan and UK, 1980, 1981, 1982)

Kimitsu works (April 1982)	*Number*	*%*	*Yawata works (April 1980)*	*Number*	*%*	*Scunthorpe works (March 1981)*	Number	%
Management	106	0.49	Management	150	0.47	Management	146	1.2
Technical, clerical	1,812	8.4	Technical, clerical	3,101	9.7	Technical, clerical	3,280	27.3
Sub-contract staff	2,112	9.9	Sub-contract staff	2,205	6.9			
Sub-total	4,030	18.8		5,456	17.0		3,426	28.5
Manual workers			Manual workers			Manual workers		
Honkō	5,440	25.4	*Honkō*	13,950	43.7			
Sub-contract workers	11,970	55.8	Sub-contract workers	12,492	39.2	Process operatives	3,132	26.1
						Maintenance and craftsmen	4,124	34.3
						Ancillary	1,324	11.0
Sub-total	17,410	81.2		26,442	82.9		8,580	71.4
Total	21,440	100.0		31,898	100.0		12,006	100.0

Sources: Kimitsu Works (Nippon Steel) 1985; Scunthorpe Works (BSC): Interview

Notes: (a) *Honkō* means production workers of Nippon Steel Corporation

(b) Kimitsu: steel output 5.8 million tonnes (1982), steel products 5.5 million tonnes (1982)
Yawata: steel output 6.15 million tonnes (1979), steel products 5.9 million tonnes (1979)
Scunthorpe: steel output 3.0 million tonnes (1979)

(c) Management includes the position of section chiefs and above (Kimitsu and Yawata) and red card and above (Scunthorpe) more managerial employees are required at establishment level. There are two main reasons for the decentralisation of management in the BSC besides cultural difference. One is that the BSC is a nationalised enterprise integrating fourteen already established enterprises; each major establishment had its own managerial employees and it was difficult to reduce this section of the existing workforce. It is not comfortable for managers to remove their subordinate managers, and it may indeed induce a sense of insecurity in themselves. The other reason is assumed to come from the difference in the size of demand, which influenced the automation in the BSC from that of the Nippon Steel Corporation, which uses automation as an integrated information system centred at head office.

The percentage of technical and clerical employees at Scunthorpe is also higher than at Yawata and even more so than at Kimitsu. This can be partly explained by the fact that a larger percentage of managerial employees naturally needs more technical and clerical employees; partly because a large percentage of jobs done by the technical and clerical employees in this category are also sub-contracted in Japan, as seen in Table 11.10; and also because a quite large number of white-collar employees are still engaged in quality control activities in Scunthorpe, whereas in Yawata and Kimitsu their duties are included among those of blue-collar workers.

Table 11.11 Comparison of workforce composition of representative large-scale iron and steel works (Japan and UK, 1971, 1972)

Yawata works (1971)			*Scunthorpe works (1972)*		
	Number	*%*		*Number*	*%*
Management	168	0.3	Management	178	0.9
Technical, clerical	4,660	9.2	Technical, clerical	4,338	23.4
Sub-contract staff	3,620	7.1			
Sub-total	8,448	16.6		4,516	24.3
Manual workers			Manual workers		
Honkō	21,704	42.8	Process operatives	6,471	34.5
Sub-contract workers	20,514	40.5	Maintenance and craftsmen	4,995	26.4
			Ancillary	2,776	14.8
Sub-total	42,218	83.3		14,202	75.7
Total	50,666	100.0		18,768	100.0

Sources: Nippon Steel 1980: 14, 33,156; Scunthorpe works (BSC): Interview
Notes: (a) *Honkō* means manual workers of Nippon Steel Corporation
(b) Yawata: steel output 7.4 million tonnes (1971)
(c) The number of management at Scunthorpe Works is as of 1974

The percentage of process operatives in the Scunthorpe works is similar to the *honkō*, the regular production employees of Nippon Steel Corporation, of the Kimitsu works, while that of the Yawata works is quite high. We can assume that it is because Yawata produces many more kinds of steel products, which require more production lines in rolling. But a more important reason is that they could not reduce the ratio of the *honkō* workers because of the established industrial relations at the works. The power of the union did not permit the management to raise the ratio of sub-contracting up to the level of Kimitsu at the expense of *honkō*. This is supported by the evidence of a lower ratio of sub-contracting in the other existing establishments, as was shown in Table 5.2, and it reflects the greater power of the labour union in the older works.

Thus we can conclude that the similar size and composition of the workforce at the establishment level is also a reflection of technology adopted at the establishment level, but their detailed differences can be explained by restraining factors active at the corporation level. They comprise historical factors unique to their developments, management strategy, the type of structuring information system (the level and combination of process and office automation) and the nature of industrial relations. For example, the percentage of process operatives seems to be determined mainly by the technological principle, as shown by the Scunthorpe percentage being almost the same as that of Kimitsu, where the *honkō* workers are almost exclusively main-line process operatives. In terms of technical and clerical employees, if a large percentage of these jobs in Scunthorpe was transferred to head office and/or sub-contracted, then the percentage of management, technical and clerical employees would become similar to that of the Yawata works. But this is mainly dependent upon managerial strategy and the situation of industrial relations, such as whether they take on a strong head office centralisation policy, or are able to introduce large-scale sub-contracting or introduce more labour-saving automation in order to rationalise their office work.

The comparison of these two works concerning workforce composition in a certain time-period, as well as a comparison of the changes observed between Japan and the UK, can be summarised as a drastic decrease in the total number of the workforce, in particular that of manual workers; an increase in the percentage of managerial employees; and an increase in the percentage of technical and clerical employees. The different trends found between the two works are that in Yawata a slight increase in *honkō* occurred, which caused a slight decrease in the percentage of sub-contract workers. This is because rationalisation of the workforce was carried out by the sacrifice of sub-contract workers in order to maintain the employment security of *honkō*. Thus *honkō* workers were transferred into the former work area of sub-contracting. This indicates how sub-contract workers are used as a flexible workforce as a buffer to changes in economic conditions.

In the BSC the absolute and relative decrease of process operatives is large, while maintenance and craftspeople increased relatively; also a slight relative decrease in the ancillary employees can be observed. We can therefore sum up the general impact of the plant and equipment modernisation on the workforce in the Scunthorpe works as follows:

1 relative increase in managerial employees;
2 relative increase in technical and clerical employees;
3 relative decrease in process operatives;
4 relative increase in maintenance and craftsmen;
5 relative decrease in ancillary employees.

Although the above common trends can be explained as a general reflection of a basically similar application of plant and equipment modernisation, subtle differences in the changes of percentage in each occupational category in comparison with the Japanese case can be accounted for by the type of modernisation, the level and different use of automation and also the strength of managerial authority to use the whole workforce, including sub-contract workers, for running the establishment as efficiently as possible.

The workforce at the corporation

Workforce composition at the corporation level reflects more the concept of corporate organisation unique to each stage of industrialisation. In terms of centralisation at head office, the extensive use of sub-contract workers and the hierarchically structured information system, Japanese companies seem to have had an advantage in organisational efficiency, especially when the demand was burgeoning. In the BSC the similarly changing trend in the increase of white-collar employees and the change in composition helped to influence the views and attitudes of trades unions towards co-operative industrial relations.

The number of employees at the head office of the Nippon Steel Corporation is much larger than at the BSC (Table 11.12). This is because of a large concentration of functions at head office in the

Table 11.12 Number of employees by head office and other establishments (Nippon Steel Corporation and the BSC)

	Nippon Steel Corporation				*BSC*			
	1960		*1980*		*1972*		*1981*	
	Number	*%*	*Number*	*%*	*Number*	*%*	*Number*	*%*
Head office	1,371	(3.4)	6,895	(9.6)	2,380	(1.2)	4,325	(4.4)
Other establishments	39,089	(96.6)	64,722	(90.4)	191,382	(98.8)	94,525	(95.6)
Total	40,460	(100.0)	71,617	(100.0)	193,762	(100.0)	98,854	(100.0)

Sources: Nippon Steel 1981; Tables 7.1, 10.3

Nippon Steel Corporation. A greater concentration of functions and power at head office and fewer managerial functions and power at the establishments were consciously pursued during the course of modernisation. In contrast with this Japanese experience, the BSC has far fewer functions at head office and they are more vertically decentralised to each managerial board at the establishment level. Although the percentage of employees at both head offices increased during the modernisation period, this difference in the trend of centralisation and decentralisation is not unique to the Japanese and British steel enterprises, it is a common feature observed among other enterprises in these countries. However, this gap is beginning to narrow as the runaway Japanese growth effectively terminated in the 1970s. Japanese steel companies began to diversify their business and move towards decentralisation in the 1980s.

In terms of the corporate culture of both industries, in general, Japanese organisations have a disposition towards a more centralised and collective bureaucratic hierarchy due to the particular influence of German advisors as early as 1926 in the Yawata works (Nippon Steel 1980: 61) in addition to the general nature of the Japanese culture, which was formed in the context of its own history. British organisations, on the other hand, have a disposition for a more decentralised, individualistic bureaucratic hierarchy, reflecting the stronger individualism among those who constitute the management class. The unique history of the nationalisation of fourteen companies into one corporation should have reaffirmed the tendency of this decentralisation in terms of corporate structure.[2] The level and scale of automation has some relevance to the corporate structure as well. The fuller use of online computer systems in the Nippon Steel Corporation has technically made it possible to absorb various managerial functions previously located at works level to head office.

White-collar and blue-collar employees

The percentage of white-collar employees in the two big steel corporations is almost the same, but slightly higher for the BSC, as shown in Tables 11.7 and 11.8. The trend of change was towards an increase of white-collar employees in both corporations. To what extent it will increase in the future depends upon the level of automation in both the production process and the office. It is important to note that although the total amount of employment is different, the ratio between white- and blue-collar employees in the two corporations is almost identical. This must however be slightly modified when we include the number of sub-contract employees, which can also be broken down into white- and blue-collar employees, as shown in Table 6.8 (p. 109). For example, the ratio between white- and blue-collar employees including sub-contracted staff was 24.1 : 75.9 in 1980 for the Nippon Steel Corporation (calculated from the figures in Table 6.8) – thus showing a lower ratio in the use of a white-collar workforce. Even allowing for the decrease in the percentage of white-collar employees when we include the sub-contracted employees in the Nippon Steel Corporation we can say that the two corporations, which have similar production systems, do need a similar combination of workforces. A strong incentive for such change is assumed to be the progress of similar mechanisation and automation in the production process and office administration and the strategy of management for the efficient use of the workforce.

This change in the ratio certainly has an important impact upon trades union organisations and their views and attitudes. An increase in the size of white-collar membership in the case of the Nippon Steel Corporation and a similar increase in the membership of white-collar workers in the traditional blue-collar unions, as well as the increased power of white-collar unions in the BSC, may influence the views and attitudes of the traditional class-conscious trades unions, thus creating favourable conditions for management to negotiate with unions for 'harmonious' industrial relations. Many white-collar employees do not identify themselves with the traditional working class even though the working conditions are getting much closer to blue-collar conditions than they used to be. In addition, the experience of white-collar employees in union activities is limited both in depth and scope. Among white-collar employees, the percentage of female employees is also almost identical between the two corporations. This suggests that the kinds of jobs for females in these two corporations are similar and they are limited in the jobs they are assigned to, such as supervisors, technicians, clerical employees, typists, machine/telephone operators etc.

Table 11.13 Composition of employees by white-collar and blue-collar employment (BSC)

	1972		*1979*		*1981*	
	Number	*%*	*Number*	*%*	*Number*	*%*
White-collar	47,322	26.9	41,212	30.5	32,271	32.6
	(8,057)	(17.0)	(7,280)	(17.6)	(n/a)	(n/a)
Blue-collar	128,158	73.1	93,615	69.5	66,583	67.4
Total	175,480	100.0	134,827	100.0	98,854	100.0

Source: BSC (Head Office): Interview

Notes: (a) Figures include only the number of employees engaged in the iron and steel activity of the BSC

(b) ( ) indicates the number and percentage of females among white-collar employees

(c) White-collar includes managerial, technical and clerical employees, while blue-collar includes process operatives, craft and maintenance employees and ancillary employees

Table 11.14 Composition of employees by white-collar and blue-collar employment (Nippon Steel Corporation)

	1960		*1970*		*1980*	
	Number	*%*	*Number*	*%*	*Number*	*%*
White-collar	13,449	19.8	22,229	27.1	22,870	31.9
	(2,560)	(19.0)	(3,884)	(17.5)	(4,049)	(17.7)
Blue-collar	54,494	80.2	59,647	72.9	48,747	68.1
Total	67,943	100.0	81,876	100.0	71,617	100.0

Source: Nippon Steel 1981

Notes: (a) The number of employees for 1960 is the combined figure of Yawata and Fuji Iron and Steel Corporations

(b) ( ) indicates the number and percentage of female employees among white-collar employees

(c) White-collar includes managerial, technical and clerical employees, while blue-collar includes process operatives, craft and maintenance employees and ancillary employees

CONCLUSION

No single factor can be adduced to explain the similarities and differences identified in management and labour in the two iron and steel industries. They arise, of course, from a combination of factors, some reflecting stronger technological influence, and others deriving more from the continuing processes of industrialisation itself. Whatever more or less direct influences may be cited to account for the present circumstances of management and labour, in the final analysis it is the dynamics of the current players – management itself, unions and sometimes government – which determine the situation.

To make a generalised comparative assessment, however, it can be observed that Japan's situation was a phenomenon appearing at a stage of organic compatibility in management, while Britain found itself in a context of contrived compatibility. Discussions in the next chapter will show how the contents of this chapter may be explained in the framework of relative advance and convergence.

12 Implications of convergence and relative advance

So far, the changing patterns and trends of management and labour in the Japanese and British steel industries have been examined and interpreted as showing some element of convergence, both of the British Steel Corporation towards the Japanese production model and of the Japanese industry towards the British situation. However, important differences remain between the two cases.

The changing trends of management and labour can be classified into three types of convergence and non-convergence, according to the direction of movement; here the causes and implications of these will be discussed in the light of the data revealed in this research work.

IMPLICATIONS OF TECHNOLOGY

Marked differences in capital investment between Japan and the UK existed throughout most of the period 1960–80, as was summarised in Chapter 11. However, the disparity has lessened following the onset of decline in the Japanese steel industry in the latter half of the 1970s. In Japan the rate of increase in plant and equipment investment slumped to almost zero in the period 1976–80 and to only 2.3 per cent in the period 1981–5. In the UK, investment increased up to 1980, but then dropped by 95.8 per cent.

Thus in the timing of investment Japan took the lead, with a level of investment much higher than that of the UK, but in the latter half of the 1970s UK investment increased greatly and plant and equipment modernisation moved ahead. This trend of investment laid the foundation for a situation of relative advance for Japan but at the same time the British investment effort to catch up with Japan has provided the conditions for convergence of the UK to Japan, i.e. the UK narrowed the difference in the level of modernisation which had previously prevailed. This is seen as an effort to maintain compatibility in management at as high a level as possible and avoid further slippage. Since the 1980s the timing and trend of investment has become relatively similar.

Because of the difference in the growth of these two national economies, the Japanese major steel corporations, unlike those of the UK, were able to establish many green-field works. Nonetheless the changing trends in specification of plant and equipment modernisation were similar in many respects in both countries, except for the quantitative aspects of equipment and their detailed arrangements. From the examination of the content of plant and equipment modernisation in Table 11.1, the common trend was to build new large blast furnaces, introduce LD converters of enlarged size, introduce continuous casters, introduce and increase new types of hot and cold strip mills and furnish these plants and equipment with advanced automation. In short, it was an effort to increase technical efficiency in the production process as well as to improve the quality of products. Thus, in terms of the content of modernisation, both countries underwent a similar process, different only in that the Japanese major green-field works have benefited from a more integrated and comprehensive modernisation with the full use of an on-line automation system under the stronger control of head office.

It must be borne in mind that there is always a tendency to narrow technological discrepancies as the pressures of competition increase, and that inevitably some time lag occurs. For example, in the 1960s Japanese steel companies imported advanced technology from Western countries, while in the 1970s Japan exported her improved iron and steel technology to both advanced and newly industrialising countries (for instance, the Nippon Steel Corporation exported some technology to the BSC's Redcar blast furnace).

Detailed arrangements in technology

As was summarised in Table 11.2 (pp. 231–3, more similarity was observed in plant and equipment modernisation between the Yawata and Scunthorpe works than with the Kimitsu works. Until a mid-term large-scale rationalisation plan was announced in February 1987 by the Nippon Steel Corporation, there seemed to have been no change in the trend of modernisation. However, with the reduced demand forecast of 90 million tonnes or less for the whole Japanese iron and steel industry, the Nippon Steel Corporation announced her own rationalisation plan, according to which the Yawata works would have only one blast furnace and cease to produce plate products. Greater specialisation towards stainless steel products is planned for Yawata. In the whole production rearrangement of the Nippon Steel Corporation, the production of iron will be concentrated at Yawata (one blast furnace), Nagoya (two blast furnaces), Kimitsu (three blast furnaces) and Oita (two blast furnaces), while three old works (Muroran, Kamaishi, Hirohata) and one new small works (Sakai) will be supplied with billets and slabs from the nearest large-scale works.

Thus, to cope with the decrease in demand the Nippon Steel Corporation aims at further specialisation of each establishment into particular products, as the BSC has already done. This indicates convergence of the Nippon Steel Corporation towards the BSC in terms of specialisation of production in each works. Accordingly there will be a thorough rearrangement of the whole Yawata works in order to set up a production system with only one blast furnace, while on the other hand in the Kimitsu works the existing four blast furnaces will continue to operate at a higher capacity and supply billets and slabs to the Muroran and the Kamaishi works, which will cease iron and steel production and specialise in rolling operations only.

The differences in the type of modernisation between the two old works of Yawata and Scunthorpe lessened even further. In the 1980s the whole arrangement of several works in the Nippon Steel Corporation needed to be reorganised in a similar style to that of the BSC, namely towards greater product specialisation at each works. Thus it has been the green-field large-scale works which were the source of salient differences in the modernisation of plant and equipment. This rearrangement of production will necessarily lead to a restructuring of management itself into a profit-based decentralised system as practised in the BSC.

CHANGING TRENDS IN MANAGEMENT

This section considers the changes in management in terms of convergence. A good number of changes in the BSC can be identified as convergence towards Japan, while movement in the opposite direction and cases of non-convergence can also be found. In the main, however, these institutional changes can be regarded as a sign of the increasing role of professional management, the establishment of harmonious industrial relations and in consequence an efficient use of the workforce.

Top management

Until 1980 the top management structure of the BSC was characterised by a clear distinction between the decision-making and executive functions, and an important changing trend since then has been the steady erosion of that division. The percentage of full-time board members from within the Corporation gradually increased to 40 per cent (1984) and all the full-time board members except one (managing director, personnel and social policy) are representatives of line functions who were not allowed to become board members in the 1970s. Thus there is a clear change in the 1980s in the composition of board membership towards stronger influence of those from within the Corporation and also those responsible for the line function. In the executive organisation, which is the second level of top management organisation, there is also a changing trend in that more people are in executive managing positions from line functions (the group and business organisations) than from functional positions at head office.

The appointment of Mr Scholey as chairman of the Corporation in April 1986 was effectively an announcement that the BSC is now in the hands of 'home grown' management and that the same people are now in charge of both decision-making and executive duties. The board is chaired by the chief executive and the people responsible for the line functions seem to be more influential than those representing the functions of head office, if their numbers are a reliable guide. Management has thus become production-oriented. At the head office of the Nippon Steel Corporation, too, people from the line have greater authority and sway than the ordinary managing directors in charge of the functions of head office. All the heads of large-scale integrated works are in vice presidential posts in the top management.

Such a clear sign of the stronger influence of internally promoted executives and the conversely weaker influence of non-executive board members in the BSC gives us evidence of convergence to the Japanese situation, and it is also a move towards a greater role for professional management. The top management of the BSC is now more independent and responsible for their business and they can plan their business policies and industrial relations with a long-term perspective and on their own enterprise basis. This again changed to the traditional British-style, importing top management from outside for a while, after privatisation in 1988, but in 1993 Mr Moffat, who had been internally promoted to chief executive, was elected chairman of the board.

Managerial hierarchy

The basic concept of management of the Nippon Steel Corporation is through collective and participative means of achieving corporate objectives at various levels, while that of the BSC is an individualistic and impositional style. The format is one of control over the jobs and individual personnel they are responsible for. Even though the basic concept and style of these two bureaucratic hierarchies do not seem to be converging from either side, there are signs of changes in both which might be seen as their management styles developing some aspects in common. For instance, the BSC has been endeavouring since 1980 to create a more flexible style of management in order to mitigate rigidity in the demarcation of functions and roles of managerial employees. This policy is found in the following aims put forward within the organisation:

1 to minimise the number of organisational levels;
2 to develop 'all-round' managers and reduce functional dependence;
3 to report through the line;
4 to use a taskforce approach on major issues;
5 to re-appraise functional committees,
6 to share services ('knock down castle walls').

(BSC: Interview)

If these attempts to improve flexibility and efficiency in the managerial hierarchy are realised, then there is a possibility that even within the same job-based contract employment there will be substantial change which will lead eventually to further developments in the institutional arrangements pertaining to employment practices.

Meanwhile, in Japan, an increased sense of equality and individualism, although based upon collective and participative values, is likely to instigate a trend towards more individualistic elements within the existing employment system, effecting change in addition to the already realised modification of the seniority-based salary structure, which includes an element of individual job and job-achieving capacity (*Shokunō Shikaku Seidō*, as a part of *Nōryokushugi Kanri*). If such changes in both managerial organisations proceed further, there will be a move on both sides to change their styles of management, thus creating similarities in their managerial hierarchies according to the incentive of increased organisational efficiency by harmonising individual with collective incentive in the UK, and in Japan harmonising collective with individual incentive.

Direct and indirect departments

Chapter 11 drew attention to the drastic change in the ratio of managerial employees between direct (line) and indirect (functional) departments of the Scunthorpe works between 1974 and 1981, which resulted in a convergence towards the practices of Japanese works such as Yawata and Kimitsu. This change suggests that such convergence is closely linked with and determined by the level of plant and equipment modernisation, which in turn has implications for increased efficiency in the organisation of the production system.

Management and industrial relations

Throughout the period of high economic growth, in-firm education and training was an important feature, as well as a means, of labour management in the Japanese large-scale steel corporations, serving to improve the quality of labour and to create advantageous industrial relations. Within the enterprise, the day-to-day conditions of the external labour market could not affect personnel management, as the management had maintained its managerial prerogative for the allocation of the workforce under its life-time employment practice.

The objectives of in-firm education and training have covered four areas:

1 to provide technical education and training based upon employees' school careers, to be carried out within the hierarchy and seen as an investment rather than an expense;
2 to foster loyalty and attitudes commensurate to higher commitment;
3 to control the promotion and salary system;
4 to bring about stronger labour control by providing a special emphasis on education and training to section chiefs and foremen.

All these roles of in-firm education and training became increasingly necessary and important as higher levels of automation transformed traditional skills, which had until then provided workers with some control over their work. As the programme and assessment of such education and training are entirely in the hands of management, the development of in-firm education and training implied a transfer of power at the workplace from workers to management.

Change has occurred here in that while the first two concerns were emphasised initially, during the high-growth 1960s, the other two areas came to the fore later, in the lower growth period of the 1970s. In addition, since 1980, when the major steel corporations began to plan their rationalisation programmes, in-firm education and training has begun to be used to retrain employees in order to dispatch them to related companies, sub-contract companies and newly-established companies which were part of their business diversification. The system of the *Nōryokushugi Kanri* (see Figure 7.2) has been especially helpful as a means of controlling promotion and reducing labour costs, and also as a tool for personnel management. In this system a modification of the existing seniority-based salary structure is achieved by establishing hierarchical qualification titles which are linked to the salary element in order to reduce total labour costs and control the promotion system. As we have seen, some modification of these traditional features of Japanese employment was rendered necessary in the 1970s with the termination of organisational expansion, and further rationalisation was needed in the 1980s. In fact, in 1981 the percentage of *Shokumukyu* among manual employees (high school graduates) accounted for 58.4 per cent, while the percentage of *Shokunōkyu* among white-collar employees (university graduates) accounted for 52.7 per cent in the Nippon Steel Corporation. This modification, emphasising individual job and job-achieving capacity, can be seen as evidence of the sort of changes occurring in Japan that bring it closer to Western employment practices.

Thus we can observe that the objectives of the in-firm education and training systems shifted more towards the control of employees' promotion and salary (labour costs) and retraining of employees for permanent transfer (redeployment of employees in various related companies), which suggest partial modification of the life-time employment practice. Such functions of in-firm education and training have been without doubt an important factor of labour management and industrial relations. They have been closely related to the salary and promotion system, the provision of corporate loyalty and intellectual stimulus, the elimination of mental strains which arise from regular routine work and the creation of a flexible working system and job enlargement.

In the BSC, the main modes of management have been to use the market mechanism, albeit restrained to a great degree by negotiations between trades unions and management, to determine the level of wages, as documented in the analysis in Chapters 9 and 10; and to use industrial relations management as a tool of labour management over employees who are on a job-rate contract and belong to various unions. Recent major changes are found in two areas.

First, industrial relations management still remains an important element and it aims to rationalise the number of negotiating parties at the corporation level; where possible a single-union industrial relations system is preferred by management as well as by influential unions such as the ISTC. Second, delegation of greater managerial authority to the divisions has taken place, concerning various important matters, including the number and structuring of jobs. As to change in the number of negotiating parties at national level, negotiations concerning general reviews of pay are now conducted with only four groups: the ISTC, the Coke and Iron Section of the ISTC (the former blast furnace union), the NJC (a National Joint Council comprised of Craft Unions – GMBATU and T&GWU), which have negotiated together since 1983, and SIMA (Steel Industry Management Association). Other conditions of employment, such as hours of work, holidays and sick pay are negotiated with the TUC Steel Committee for industrial grades and staff grades, and with SIMA for management. Besides negotiations, joint consultations are also increasingly conducted at national, works and departmental levels with the TUC Steel Committee and SIMA in order to create better communication and understanding between management and labour. Thus a trend towards an enterprise-based negotiating and consulting process, with fewer trades unions, has been evolving, which has significantly helped management improve the efficiency of the negotiation and consultation process as well as created a more participative climate.

As to the second change, the introduction of LSB (Local Lump Sum Bonus Scheme) and the subsequent developments as a result of its success should be noted. The establishment of this scheme was a major point at issue in the 1980 steel strike, but the management has succeeded in establishing it as a strong tool for reductions in cost and improvement of industrial relations at the establishment level. Various changes are in progress which will transform the traditional relationships established between the management and trades unions. Some of these are given below, even though they are still evolving.

1 The traditional process grades wage structure, established upon seniority-based promotion lines by trade union regulation, is being challenged by management. Over recent years management have insisted that many critical jobs should be on a selection-only basis.
2 The rejection of traditional departmental and trade union demarcations, is creating an attitude of co-operation.
3 The introduction of a 'team working' approach will minimise the workforce, using a flexible working system.
4 Labour-saving automation has been introduced.
5 There is an increased use of sub-contracting in services and for irregular peaks, wherever it is cost-effective.

All these changes emerging at the workplace under the initiative of management manifest a strong move towards a new model of industrial relations and show signs of convergence towards Japan. It is not, however, correct to claim that all industrial relations will proceed to a common model in the near future, for their nature is relevant to the historical process of industrialisation and it differs from country to country. If the essence of industrial relations still lies in the relationship between capital and labour, the recent trend is a more compliant attitude of labour under the logic of capital accumulation. This situation is likely to accelerate in the 1990s, as we shall see in the final chapter.

Workplace management

The changing trends in workplace management and the role of the foreman will give us an understanding of convergence and non-convergence in the most important section of the works. Workplace management and the foreman's role in the 1980s are different in the two countries, but in general the BSC seems to be approaching Japan in terms of the emergence of new work practices and co-operative industrial relations. The Japanese steel corporations have shown a unique way of dealing with the large-scale rationalisation of the workforce, supported by a well-established workplace management and a management-responsible role for the foreman, while in the BSC a move towards good industrial relations has introduced various institutional changes which have replaced the old institutions in the midst of the large-scale rationalisation of the workforce.

In the 1980s there were no special changes in the style of workplace management and role of the foreman in the major Japanese iron and steel corporations, but there was large-scale rationalisation aimed at reducing production capacity and as a consequence staffing, particularly in the older works. There was no serious workplace conflict hindering the rationalisation plans, thanks to the co-operative nature of industrial relations. The union attitude is to demand the minimum reduction of workforce and to create alternative opportunities for employment for those who have to leave the company either temporarily or permanently. Such employees are provided, through negotiation with the enterprise union, with jobs in sub-contract companies, affiliated companies, steel-consuming companies (e.g. automobile firms) and companies newly-established by the steel corporations. The Nippon Steel Corporation dispatched, rather than dismissed, 785 employees in 1986, and they went mainly to vehicle manufacturers, including Toyota, Daihatsu, Mitsubishi and Suzuki. Thirty-two companies have been set up by the Nippon Steel Corporation, creating 2,053 new jobs in the period 1980–7. Further measures for rationalisation are:

1 to suspend the extension of the retirement age, which was originally planned to be extended from 55 to 60 years of age;

2 to provide a favourable retirement lump sum bonus for those who retire early;
3 to provide special education and training for those who are transferred to other workplaces or works in the same enterprise, and also to other companies, which represents a permanent dismissal.

The above measures for rationalisation are fully supported by the union and most of its members. Some changes directly relevant to workplace management were also made:

- to encourage foremen over 55 years old to abandon their position and revert to the rank of ordinary worker in order to give younger workers a chance to become foremen. A foreman who gives up his position will continue to receive the same salary for another six months before his foreman's allowance is cut off. By this measure the management expects to provide incentives to younger workers for promotion to the foreman position and at the same time to reduce the labour costs which accrue from the seniority system.
- to change the salary structure to one which includes an increased percentage of elements determined by evaluation of job and job-achieving capacity, while reducing the percentage of the seniority-based element. By this increased percentage of merit-rating in the salary structure the management is able to increase its managerial prerogative.

What is perhaps unique to the Japanese steel corporations is that such changes are common both in timing and content to all the corporations in the industry. Such rationalisation has created a partial modification of life-time employment in the steel corporations, but the principle of life-time employment and its related system of labour management are maintained for those employees remaining in their parent company. However, those moved into other companies have similar labour management but perhaps poorer working conditions and lower salaries. Thus the advantage of the so-called good industrial relations in the Japanese large steel companies is that they allow management to deal smoothly and flexibly with changes in external conditions.

In the BSC, although some similar measures to cope with the rationalisation of the workforce are found, there has been one unique change, as explained in the preceding section, relevant to workplace management and the role of the foreman. One recent example of such a change is the introduction and implementation of tubemaster teams and multi-skilled craftsmen at the Corby Tube works in 1987. The agreement was made between the BSC Corby Tube Works and the

ISTC, the ISTC (Coke and Iron Section), the AEU (Engineering Section), the AEU (Supervisory), the AEU (Technical, Administrative and Supervisory Section) and the EETPU with regard to increase in pay and other matters. This new attempt at the Corby works is set to undermine the existing industrial relations. Some of the important features included in the above agreement are as follows:

- there are only two grades of employees, a team leader and team members and existing occupations will be progressively phased out;
- each team is responsible and accountable for all its activities of production which include quality, cost, output and yield;
- each team member will undertake any production or maintenance task with unrestricted flexibility and mobility;
- ISTC staff supervisory occupations will be discontinued and the job holder will revert to manual grade conditions with the manual grade wage structure;
- multi-skilled craftsmen undertake any maintenance activity either mechanical or electrical, thus eliminating job demarcations between the existing mechanical and electrical craftsmen and providing an effective incentive to craftsmen.

These changes suggest the following implications for changing industrial relations in the BSC:

- creation of a group working system and elimination of existing occupations will inevitably induce an erosion of union seniority and the introduction of a new wage and promotion structure based upon a grading system using job evaluation specific to the conditions of the BSC;
- team-work concepts and responsibility for total quality performance will create a sense of participation and co-operation for achieving a team/organisational objective;
- team members undertaking the tasks of operation and maintenance in the area of the major production process suggests that this major production process will be under the control and responsibility of ISTC members alone. This will create the possibility of using the workforce in a similar way to the Japanese companies which are using only *honkō* employees in this major production process, while sub-contracting jobs in the maintenance and peripheral areas;
- dissolution of the foreman's functions into the lower ranks of middle management and group leaders is a unique process which has developed in the historical process of industrial relations in the UK. However, if branch officers of the ISTC together with leading union

officials negotiate with managers of the BSC over labour management issues, and the jobs of work management are absorbed by middle management, then it becomes a rationalisation of industrial relations; this constitutes an important change in the nature of industrial relations at the BSC;

- integration of the mechanical and electrical occupations will reduce the demarcation disputes and create conditions for good industrial relations at the workplace; in particular, if the various craftsmen come to be organised into fewer trades unions, then industrial relations with the NCCC (National Craftsmen's Co-ordinating Committee) will also become more harmonious, although this will not proceed quickly and smoothly as the members of the NCCC are workers with a strong working class consciousness.

Thus there are various signs of convergence towards Japan in the workplace management of the BSC and they can be regarded as changes towards a function similar to that found in enterprise-based industrial relations. But a clear non-convergence trend, as seen in the different role of the foreman in the two corporations, shows that historical influence will not easily disappear, but rather exert a continuing influence upon the new industrial relations institutions.

CHANGING TRENDS IN LABOUR

Labour at the workplace

The size of the workforce at the workplace is always influenced by the technology adopted by management. We can confirm that similar types of equipment tend to result in similar numbers of workers, as was examined in Chapter 11 concerning the number of workers at workplaces of similar capacity blast furnaces, such as Kure and Scunthorpe. At the same time the modernisation of plant and equipment tends to reduce the relative number of workers operating that equipment. For example, the comparison of the number of workers between a modern new blast furnace at the Kakogawa works (see Tables 6.11 and 6.12) of Kobe Steel Ltd and the old blast furnace at the Kure works of Nisshin Steel Co. Ltd (see Table 11.9) gives us evidence that a relatively smaller number of workers is needed for modern large-scale equipment. The Kakogawa works (No.1 and No.2 B.F.) has only twenty-two workers at the blast furnace workplace for an annual capacity of 5,837,000 tonnes (265,318 tonnes per person), while the Kure works (No.1 and No.2 B.F.) has nineteen workers for 3,059,000 tonnes (161,000 tonnes per person).

As to the average age and length of service at the blast furnace workplace, a comparison between Kure and Scunthorpe (see Table 11.9) shows that there is broadly a similarity in age composition, even though the length of service is much shorter at the Scunthorpe works. This suggests two things. One is that the rate of turn-over is higher as a reflection of employment practices in the Scunthorpe works, and the other is that in modern plant and equipment the traditional experience-based skills are not as strongly required for operatives as they used to be. The Kimitsu works of the Nippon Steel Corporation provides us with more evidence. The average age and length of service in the Kimitsu works in 1971 were 29.6 years old and 9.8 years respectively, but we can assume that when this works was opened in 1961 it started operations with workers who were much younger and whose length of service was much shorter.

Although age and length of service are not the same among Japanese and British workers and also a larger difference exists between new and old works in Japan, we can recognise through the examination made in Chapters 6 and 9 that management has attempted to reduce the average age and length of service both in Japan and the UK. This has been for various reasons, but mainly to reduce labour costs and create good industrial relations. It was a more acute situation for the management of Japanese corporations, where a large percentage of salary (about 50 per cent) is determined by seniority grading scales. Thus there is a strong case for modifying the salary structure without losing the sense of loyalty among employees, in order to reduce labour costs. Thus, the number of workers and related trends can be seen as almost immediate manifestations of technology, although specific age and length of service of workers can be differentiated, influenced by managerial strategy, employment practice and industrial relations.

Labour at the establishment

The effect of plant and equipment modernisation at the establishment level is first confirmed in the drastic decrease in the total number of workforce, intended to reduce labour costs. At both Yawata and Scunthorpe works the size of the workforce decreased drastically in 1971/72 and 1980/81. Considering the number of workers at the most modern Kimitsu works, it is likely that the workforces at both Yawata and Scunthorpe will fall further.

Second, the percentage of management, technical and clerical employees has risen in both works during 1971/72 and 1980/81. This trend has been seen as a common consequence of plant and equipment modernisation. But if we look at the most modern Kimitsu works, the percentage of technical and clerical employees of the Nippon Steel Corporation is the lowest among the three works. This suggests that with the use of the comprehensive on-line computer system controlled by head office, a large number of technical and clerical functions were absorbed into the head office management system and at the same time quite a large number of clerical and technical functions were also transferred to the white-collar employees of the sub-contract companies whose offices are in the same works.

Third, manual workers in the Scunthorpe works are divided horizontally into three categories based upon occupational classification, while in the Yawata and the Kimitsu works they are divided vertically into *honkō* (manual workers of Nippon Steel Corporation) and *shitaukekō* (sub-contract workers). The three manual categories of the Scunthorpe works reflect the history of the occupational formation of manual workers and how they have been organised by the trades unions since the nineteenth century, while the use of the workforce in Japan reflects an optimally flexible and stable use of a workforce without obstruction by the growth of a class-conscious union movement, which failed to gain a hold at the start of the late and rapid development (1950). There are however some new trends appearing in the BSC since 1980 in an attempt to re-organise the manual workforce; this is a management-led drive, with the consent of the trades unions, to cast off traditional institutions. One is a trend to use sub-contract workers and the other is to eliminate job and occupational demarcation, which will eventually aim at the creation of a steelworker concept as well as the BSC employee concept. This experience has already been reported on from Port Talbot in the 1960s (Dore 1973: 345), but the effort was not successful at that time. Many recent agreements on flexibility reached with trades unions suggest that this policy is being re-introduced, but this time more successfully.

As to the use of sub-contract labour in the BSC, it is classified into three groups: capital development construction schemes; major revenue maintenance schemes, e.g. relining/refurbishing a blast furnace; and maintenance 'peak-lopping'. The trades unions in the BSC have generally accepted the first and second sub-contracting practice, but sub-contracting in the third case has often created conflict between management and unions. Since 1980 the plan to use sub-contract labour has gone further: sub-contract labour has entered peripheral activities in most BSC plants: catering, office cleaning, general amenities, general plant cleaning, external road transport, heating and ventilating, painting and decorating. Sub-contract labour has also expanded in some mainstream activities: furnace-wrecking, refractory bricklaying, internal works transport, slab-dressing.

The use of sub-contract workers in connection with the workforce composition has important significance for institutional arrangements in terms of convergence towards Japan. It moreover provides evidence against any refutation of such convergence. If something has not occurred in the immediate period of modernisation it does not mean it will not happen in the future, and the opportunity now seems to be taking shape for the greater use of sub-contracted workers in BSC workplaces. Thus, although there are salient differences in the size and composition of the workforce at the establishment level, the changing trends at this level seem also to be common and as a result such differences seem to be narrowing.

Labour at the corporation

A notable difference is found in the workforce composition of head office and other establishments between the Nippon Steel Corporation and the BSC, and although the tendency to increase the percentage of head office employees has been similar, reflecting similar technological conditions, the difference does not seem to be lessening. This is mainly due to a difference in the basic concept of organisation structure, which is a restraining factor against technological influence. This is represented in the centralised corporate structure of the Nippon Steel Corporation and the decentralised corporate structure of the BSC. Because of this, the use of automated systems in the whole organisation seems to be quite dissimilar. Although in the 1970s there was an attempt in the BSC to have a similar centralised structure with more concentration of control functions at head office, it was decided that it did not suit the BSC because of insufficient growth in demand. Thus, this is an area where there still remains a large disparity between the two corporations. It is no surprise to note, however, that the Nippon Steel Corporation began in the 1990s to show signs of convergence in this area towards Britain, when faced with no future prospect of growth in demand, as will be shown in the next chapter.

As to the ratio of white- and blue-collar employees at the corporation level, there are some differences between the two corporations, but here too there is a strong common trend in that the ratio of white-collar employees has been increasing. At present the percentage of white-collar employees is slightly larger in the BSC, as the Nippon Steel Corporation excludes the sub-contract workers. But this difference might be narrowed if the BSC increases sub-contracting in more areas of production and some of the white-collar jobs, such as those in quality control, are transferred to manual workers. The similarity of these ratios does not tell us anything immediately about convergence in institutions, though it is relevant that the type of labour increasing (white-collar employees) affects the structure and nature of institutional arrangements, especially industrial relations.

RELATIVE ADVANCE AND CONVERGENCE

The aim of this study has been to examine the two steel industries in terms of the impact of technology upon management and labour and explain them in terms of relative advance and convergence. The principal convergence theories were concerned with industrialisation and its impact on the macro-economic and social dimensions, but the aim of the present work has been to examine it at the micro level, i.e. management and labour. An examination has been presented of the effect of plant and equipment modernisation upon management and labour, rejecting the common convergence theory that views technology as given; in addition, institutional arrangements were not here considered to change automatically with the onset of technology.

Although there have been some important differences in the timing and size of equipment modernisation and the level of automation between the two industries, the similar technological paradigm has produced similar changes in job content, a relative as well as absolute decrease of the workforce when faced with greatly reduced demand, and in workforce composition. In this process the role of professional management has increased greatly; in particular those internally promoted came to be increasingly responsible for the running of the large-scale corporations. Internally promoted professional management in Japan tended to include the corporate welfare and the well-being of employees as part of their managerial goals during the high growth period, which was an agent in creating harmonious industrial relations as well.

The BSC, which at first suffered from traditional institutional rigidity and inconsistent government policy, has begun a delayed plant and equipment modernisation and a thorough rationalisation. They have greatly reduced the workforce and its composition has also been substantially changed. Such changes in the size of the workforce and its composition have influenced the views and attitudes of trades unions and provided management with conditions to create suitable industrial relations and increase managerial authority for institutional changes. A good number of additional changes which can be taken as convergence towards Japan have also been under way since 1980. The patterns of convergence and non-convergence listed and explained in the following section give a summary of change in management and labour in terms of relative advance and convergence.

Convergence of the UK

This puts forward the view that Britain is converging towards Japan, mainly in two areas: internally promoted top management and management and labour at the workplace. The importance of internally promoted top management reflects the changing view of owners of business to make much of production, while the latter reflects the necessity of restructuring the institutional arrangements largely established in the nineteenth century and modified since then. The delay in plant and equipment modernisation of about fifteen years in the BSC compared with Japan seems to have given a technological reason determining the direction of convergence to Japan. This gives rise, of course, to the relative advance of Japan, due to Japan's earlier plant and equipment modernisation. Some of the major findings of convergence in this respect are as follows:

1 Top management function and importance have effected some convergence towards Japan.
2 Works organisation has been greatly rationalised in the area of direct and indirect organisations and it has converged to a style similar to the Japanese.
3 Management and industrial relations have been affected in that institutional changes as given below suggest signs of more participatory and co-operative industrial relations:

   (a) bargaining structure
   (b) joint consultations
   (c) participation and co-operation
   (d) introduction of sub-contractors
   (e) introduction of labour-saving automation.

4 Workplace management has changed in the following ways as a result of the above change in industrial relations:

   (a) elimination of occupational categories
   (b) introduction of team work
   (c) introduction of flexibility and mobility in the team
   (d) combination of regular operative and maintenance work
   (e) craftsmen responsible for any maintenance work.

The change in top management structure, such as the internally promoted chairman and the composition of board members, can be explained by the political influence of nationalisation. Internally promoted top management is considered to have a stronger incentive and responsibility for the success of their own enterprise, as their position and remuneration will be secured by the result of their managerial performance. This view, which supports managerial autonomy, seems to reflect the current political thinking of the government, which favours privatisation and self-financing of nationalised enterprises.

The change in management and labour at the workplace can be explained by the impact of technology and the managerial response to it in order to create an efficient production system under plant and equipment modernisation. The changes in workforce composition have influenced the trades unions' views and attitudes, providing management with an opportunity to increase managerial prerogatives against trades unions and to change industrial relations towards a management-led co-operative style. The trades unions have even agreed to introduce sub-contracting on quite a large scale. As the modernisation of the BSC post-dated that of the Nippon Steel Corporation, its impact was also post-dated. Thus the change in management and labour at the workplace can be taken as a phenomenon of dissolving occupational demarcations and the creation of multi-functional workers who can be organised under a single trade union, which suggests a sign of convergence to the Japanese model.

Convergence of Japan

This convergence provides us with evidence for the view that Japan's macro corporate environment has been converging towards that of Britain, which means that Britain is in a state of relative advance in the process of capitalist development. The direction of this convergence follows from the fact that Japan's industrialisation was late and rapid, and recently has caught up with the West to enter a stage of maturity or relative decline. Areas of convergence in this direction are as follows:

1 Introduction of *Shokunō Shikaku Seido* as a means of shifting from the seniority-based pay and promotion practice into a job-based Western model, in order to reduce labour costs and increase individual incentives.
2 A sign of overall re-arrangement of establishments within the Nippon

> Steel Corporation suggests that it is now moving towards a product-based cost or profit-centre system as found in the BSC. Its effects are especially notable in the older works. Such a rearrangement will eventually lead to some important change in corporate structure.

The introduction of *Shokunō Shikaku Seido* in 1970 can be explained as a managerial response to its own salary structure. It aims to reduce the labour-cost increases resultant from seniority-linked salary increments, which accompany the end of organisational growth. It also aims to rationalise personnel management by providing more individual incentives to employees, especially white-collar employees, whose percentage has increased as a result of plant and equipment modernisation. The major cause of the recent development of this scheme is the end of growth in demand; the corporate environment influences management to change salary structure with the aim of reducing labour costs. The strategies proposed in 1987 from the major companies were more or less similar and aimed at reducing the percentage of the basic, seniority-based, rate from 50 to 40 per cent, increasing the job rate to 35 per cent and the job capacity rate to 25 per cent. As a result, the salary of employees over 45 years old will be held back, while those over 50 years old will suffer quite a large reduction. Thus the ratio of the elements of job and individual capacity has come to account for 60 per cent, which suggests a further modification of the traditional seniority-based salary structure towards Western models.

The restructuring of establishments of the Nippon Steel Corporation in a manner comparable with the BSC can be explained by the recent sudden decrease in demand. This decline in demand is not of a temporary nature, but will continue due to the maturity of the Japanese economy and pressures from globalisation, which encourage the internationalisation of corporate activities. The large-scale steel companies are already putting emphasis on overseas investment and diversification of business activities. There will be further specialisation among establishments by concentrating iron and steel production at efficient major works and making older and smaller works specialise in the production of particular finished goods, as has taken place within the BSC.

Mutual convergence

Mutual convergence in the strict sense of managerial institutions was not found, but an important similarity in the trend of workforce composition created a condition wherein management may pursue a similar corporate strategy to reduce inefficiency in an enlarged managerial hierarchy. An attempt to increase flexibility in an enlarged managerial hierarchy has been made, although the characteristics of the two managerial hierarchies still remain broadly as before. The basic concept in the approach to reducing the rigidity of managerial hierarchy lies in the harmonisation of collective incentive with that of individuals for the Nippon Steel Corporation and that of individual incentive with that of the group for the BSC.

An institutional change aiming to increase flexibility in managerial hierarchy for the Nippon Steel Corporation demands a further modification of *Shokunō Shikaku Seido*. They intend to use it as a means of flexible control of managerial employees and to counteract rigidity among sections and departments. In the BSC, various measures aimed at mitigating rigidity in the demarcation of the functions and roles of managerial employees have already been proposed and implemented.

Thus the response of management to the rigidity of the managerial hierarchy tends to achieve compatibility in management by creating an optimum compromise between individual and collective incentives regardless of the difference in ownership and cultures of both companies.

Non-convergence

This is where no immediate change has yet appeared, but in the long term convergence may also occur as the corporate environment continues to change. The turnabout in the phase of industrialisation in Japan, from organic to contrived compatibility, will eventually influence the concepts of the Japanese corporate structure and the role of the foreman. Britain is relatively advanced in the above two areas. Therefore it is Japan that will eventually converge towards Britain. Neither technology nor macro corporate environments are effective enough at the moment, but delayed convergence may occur in future. The culture and values formulated at an organic stage of industrialisation in each economy seem to be partly responsible for this delay in convergence. The following two areas may be seen as cases of non-convergence:

1 The concept of corporate structure: centralisation of the Nippon Steel Corporation and decentralisation of the BSC.
2 The role of the foreman: higher status in Japan and a lower and diminishing role in the BSC.

The type of corporate structural shift – whether centralisation in Japan or decentralisation in the UK – may be explained by whether the corporation had been at the stage of organic compatibility until recently or at the stage of contrived compatibility for a long time. The size of demand and different histories of corporate development will also influence this. In the 1970s, the BSC attempted to introduce centralisation as part of managerial strategy to follow the Japanese model, but it was eventually abandoned for the reason that the decreased demand made a quite different approach necessary. The BSC has now returned to a more decentralised structure with restricted managerial function at head office. The Nippon Steel Corporation, on the other hand, has historically been centralised since the establishment of the company in 1897 as a government steel works. As the initial period of industrialisation in Japan was relatively shorter and the speed relatively faster than in other advanced nations, most of the large-scale companies in the key sector were encouraged and supported by the government, adopting a similar centralised structure and tending to make head offices more influential as the nerve centres of the organisations. Centralisation was thus an effective structure for companies even in the pre-war period. The post-war industrialisation of Japan recreated a similar phenomenon in the process of industrialisation, but this time on a larger scale. The major steel companies of Japan were at the stage of organic compatibility until recently. While they are at that stage, the centralisation of functions and the enlarged head office seem to be common features. Unless the scale and period of change in the corporate environment reaches a suitable level, top management may not be forced to change these concepts. Due to the corporate culture and values, top management will tend to avoid any drastic change which may influence the managerial employees who surround top management.

Finally, in the early 1990s the major Japanese steel companies began to restructure head office organisation and the corporate structure. This was around fifteen years after the reversal of compatibility from organic to contrived. As decline in demand sets in, a corporate structure once thought to be advantageous and efficient begins to seem less so; at this stage, restructuring of the head office, as well as its function, will be considered. Signs of decentralisation have already emerged in the 1990s in all major Japanese steel companies and we might take it as a sign of delayed convergence towards Britain.

The role of foreman seems an area of non-convergence, but could also be an area of delayed convergence. The relatively lower status of British foremen may, surprisingly, be taken as a case of relative advance and in future the role of Japanese foremen may show some sign of convergence towards Britain. The industrial relations of the BSC in the 1990s shows a sign of good industrial relations, but the history of class-based industrial relations still persists. The managerial function previously held by foremen seems to have been absorbed by the lower ranks of middle management, and the experience-based skills have been taken over by group leaders appointed to newly-made work teams. This declining status of the foreman could signify a more advanced pattern of industrial relations, while in Japan workplace industrial relations still depend heavily on a paternalistic foreman system as a remnant of the stage of organic compatibility. With the condition of continued low economic growth and further change in corporate environment a convergence in this area towards Britain may, as hinted in the last chapter, occur in Japan.

There are still some areas which seem unlikely to converge in either direction, due to the influence of the respective cultures of the two industries and societies. For example, although there is sufficient evidence for convergence in the area of industrial relations from the BSC to Japan, there is no sign of an employee director scheme in Japan. This piece of industrial democracy, developed in the BSC in the latter half of the 1970s, began to decline in the 1980s and was taken over by an entirely different climate of industrial relations in the 1990s, to the extent that one must wonder if any genuine co-operation between the two sides is possible in the framework of a capitalist economy. In Japan union compliance proceeds as soon as the compatibility shifts from the organic to the contrived stage. Industrial democracy in the style of the BSC is unlikely to emerge in Japan in the near future, either from inside the corporation or from any legislative influence. The collective nature of the Japanese workforce will not easily disappear either, because the Japanese have not had long enough to develop strong individualism, as the British did. Any restructuring in institutions there will emerge on the basis of existing culture and value systems.

CONCLUSION

Separating the concept of convergence into types – convergence A to B, convergence B to A, mutual convergence and non-convergence – is a valuable tool in identifying and comprehending the shifting processes of change in management and labour environments. It can be seen how situations of relative advance may emerge either through technological development or through historical process, each affecting crucial elements of the corporate development and conferring certain recognisable features.

Technologically, Japan in the post-war decades realised a situation of relative advance in production systems, initiating a movement of convergence from Britain towards Japan; conversely, the long history of industrialisation in Britain created a unique feature of corporate development which represented relative advance in comparison to Japan, which again has initiated a trend of convergence. In particular, the areas of corporate structure and the role and position of the foreman seem liable to become foci of delayed Japanese convergence towards the British situation. In addition, in some areas there was movement from each side towards a middle ground, in an effort to reduce organisational inefficiency, while other areas, for deep-rooted cultural reasons, seem at present unlikely to converge.

13 Further changes in the 1990s

In the thirty years 1960–90, the Japanese steel industry moved through a situation of organic compatibility in management to one of contrived compatibility as from the late 1970s. Throughout this period, the British steel industry was experiencing contrived compatibility, but although delayed in terms of technological innovation, its corporate structure was already relatively advanced, owing to the changes it had undergone in its corporate development. Despite its condition of contrived compatibility, the BSC established a production system similar to that in Japan in the late 1970s, and also made certain institutional adjustments, all of which resulted in convergence towards the situation of the Japanese steel industry.

This final chapter considers the most recent features of change in the context of management and labour in both Japan and Britain and assesses their implications for the theoretical assumptions employed in the present work.

CHANGE IN CORPORATE ENVIRONMENT

The average growth rate in Japan's GDP from 1990 to 1993 was 2.2 per cent, while Britain's was –0.3 per cent. These rates constitute the lowest since 1960; the growth rate of Japan has thus converged to that of the UK, and indeed Japan also had a year of negative growth in 1993. This decline in Japan's growth rate and other macro-economic conditions such as globalisation have induced the Japanese steel industry to restructure itself in order to regain compatibility in management. The recent similarity in the corporate environment of Japan with that of Britain has brought about greater convergence in the approach of management and corporate structure towards Britain, while at the same time the current political mood in Britain, which favours liberalism in the business world, encouraged the privatisation of the steel industry.

In December 1988 the BSC was privatised and since then has carried out further modifications to management and labour.

It was the start of the economic recession in April 1991, together with the rapid rise of the yen in foreign exchange markets[1] that forced the Japanese steel companies to plan a drastic restructuring in management and corporate structure. The steel output of Japan continued to decline, from 110.3 million tonnes in 1990 to 99.6 million tonnes in 1993, although the export rate of steel was maintained at a level of over 20 per cent. The UK steel output also dropped, from 17.8 million tonnes in 1990 to 16.6 million tonnes in 1993. Thus the demand trend in both Japan and the UK seems now more of a common circumstance and has become constant, although the two remain different in their scale of production. The structural change in the Japanese economy suggests that even if it should recover at all in the near future, steel demand in Japan will not increase significantly. Under this forecast for Japanese macro-economic conditions, all the major steel companies announced schemes for drastic restructuring in March 1994.

RESTRUCTURING IN JAPAN[2]

Smaller head offices

In contrast to the large head office, analysed in Chapter 7, the major companies aim to create smaller but effective head office organisations. Nippon Steel Corporation, for example, merged two top management organisations into one, namely *Keieihōshin Kaigi* (Corporate Policy Council) and *Jyōmu Kai* (Managing Directors' Council), which became *Keiei Kaigi* (Corporate Council).[3] Executive functions will be transferred to the line managers as much as possible and the line hierarchy will be rationalised into three positions: executive director, general manager and room head. In head office organisation itself the number of staff will be halved by merging twelve organisations into seven. For example, the planning, related business and general research departments will be merged into the corporate planning department, the accounting and finance departments into the finance affairs department, the general affairs and information system departments into the general affairs department, and personnel and labour into the personnel and labour department. The above rationalisation, if implemented thoroughly, will reduce the number of top management and head office employees, thus making head office much smaller in size, with a function limited to basic and corporate-level decision-making; this shows a sign of convergence towards Britain. In retrospect, we might conclude that the unitary enlargement of head office which in Chapter 7 we thought to be a unique feature of large Japanese steel companies was a phenomenon at the stage of organic compatibility, based upon corporate growth in a climate of high and continuous economic growth. Now the change in macro-economic conditions demands that the company restructure its head office and top management in both size and function.

Division system

Some Japanese steel companies had previously had engineering and other divisions, but they were not business divisions in the strict sense of the term. They were considered part of the main steelmaking activity. Nippon Steel Corporation proposals envisage the dissection of the steelmaking business into a product-based division system, producing a cost and profit analysis for each product. They also plan to establish a product-based sales and technical department, in order to meet the demands of markets. This will be, in principle, a step towards a division system like that introduced by the BSC in the late 1970s. This system will rationalise the organisation of steel works and the R & D and equipment department. The total number of departments of all the works, which now totals eighty-one, will be reduced to sixty, while thirty-four organisations associated with the R & D and equipment department will be integrated into twenty-four organisations. Thus, they aim to create a slim but efficient steel works with an organisation concept of decentralisation which increases the relative importance of line authority. For example, there is an increasing tendency that line managers have more say in the decision-making of personnel matters within their own department. Another example is that the well-known transfer of employees will also be decreased compared to the 1970s and 1980s. Instead of transfer, specialisation will become the requirement. Transfer will remain at the higher levels of management, such as for general managers. The rationale for transfer at this level is that the ability and function required at the level of general managers is more common, and also different criteria will be used at this level with more commitment from top management, in particular the president. In future it may well transpire that an independent division will begin to recruit employees independent of head office. Decentralisation has become a concept suitable to the phase of contrived compatibility in management, when it is necessary to come to terms with low economic growth as a constant corporate condition.

Diversification

Diversification of business for the steel companies became an objective of corporate strategy when the Japanese economy terminated its high growth in 1973. Until then the companies had not envisaged any other businesses being required other than the production of steel, which was sufficient in itself and still developing. In the latter part of the 1970s, and up until 1985, the emphasis of diversification in the steel companies was placed upon steel-related businesses such as overseas technical assistance and engineering. Thus diversification was an extension of, and part of, the overall steel business. In the latter part of the 1980s, however, the steel companies expanded into areas of non-steel business, such as information and communications, electronics, chemicals, new materials, bio-technology, urban developments, social services, etc. Such diversification was useful and necessary as a means of transferring employees made redundant under the practice of life-time employment, but at the same time it was not fully expected that any of them would turn into a profitable business. For example, the Nippon Steel Corporation had about 3,000 employees engaged in system engineering, and established four systems and information-related companies using these employees. They also established a large leisure park called Space World on land previously part of the Yawata steel works. The percentage of the company's income accruing from such new businesses accounted for 25 per cent in 1990 and they aim to increase this figure to 40 per cent, which would create a total composition of income by 1995 as follows: steel, chemicals, new materials 60 per cent, electronics, information, communications 20 per cent, and social and leisure developments 10 per cent.[4]

The above new businesses, if successful, will become independent companies affiliated to the parent steel companies. This type of diversification was not experienced by the BSC. The case of the Japanese diversification can be explained by two reasons: one is the pressures from the existing practice of life-time employment and the other is due to the technological advantages they created during the high growth period. The profits were spent not in the steel business, as it was already mature, but in non-steel businesses, in order to use their technological resources as well as a relatively surplus workforce.

Overseas business

Overseas investment for steel production was not considered an important business activity by the steel industry until recently, when steel-consuming industries moved their production overseas. The investment in the past had been limited and designed to ensure the procurement of raw materials such as iron ore and coal; hence they were concentrated mostly in Asia, Canada and Australia.[5] However, after a certain period of technical assistance to steel companies in advanced countries such as Britain and America, they began to invest abroad even in the business of steel production from the late 1980s. This was something the steel companies themselves had never envisaged in the past, having thought it unprofitable in terms of the labour costs in advanced countries and also an activity liable to encounter resistance from the existing steel industries in these countries. Conditions have changed in recent years, however, and nowadays steel companies tend to consider setting up production at the most appropriate place in the world in terms of market and other economic factors. The type of investment is, however, limited to joint ventures with existing steel companies and the majority of partners are in the USA. The steel companies have invested in these countries as the demand for other Japanese companies, in particular automobile manufacturers in the USA, has increased. Nippon Steel Corporations began two joint ventures with US steel companies in 1989, another two in 1990 (USA, Thailand) and one in 1993 (Thailand). Kobe Steel, which has a history of overseas operations, now has nineteen investments – ten in the USA and nine in South East Asia.[6]

This type of overseas investment can be regarded as part of the recent corporate strategy of the major steel companies, due to the maturity of the domestic market, the rising yen and the increase of overseas investment by the automobile industry. The nature of their activity can be taken as an expansion of their domestic business overseas. The BSC has overseas operations in about fifty different countries and they were usually profitable even when the BSC itself was in deficit. Overseas investment for any industry is both part of and a process of capital accumulation, and the Japanese steel industry seems to have arrived at this stage in the 1990s in the process of globalisation. Thus in terms of the process of capital accumulation the Japanese steel industry shows a sign of convergence towards what is now British Steel plc.

Employment

The common objectives of restructuring raised by all major steel companies was to maintain corporate growth and to create a strong but lean corporate structure. Such an objective was the ideal of top management, and one which is not easy to achieve; although it is on paper a sound idea from the management perspective, it presupposes both fewer employees and more work for the regular workforce. In the 1980s the common approach to redundancy was to transfer employees to other businesses created by the steel companies, while still maintaining them as employees of their parent companies. The management of the early 1990s would like to discontinue their contracts with those transferred as soon as possible, yet still maintain life-time employment for the remaining key employees who can serve in the lean production and future growth of the company. The implication of such a change in employment practice for our thesis is dual; for those who leave before the age of compulsory retirement it would mean the abandonment of life-time employment, i.e. a convergence towards Britain, while for those who remain there is still life-time employment, representing an area of non-convergence.

Life-time employment in its strictest sense has not really existed anywhere, but the idea and the predisposition to employ people in the long term existed generally in Japan during the period of high economic growth. It is rational and appropriate when companies are growing and the external labour market is tight. As the corporate environment has changed, however, this practice of long-term employment has had to be reformulated. The start of this reform began with the reconsideration and revision of the *nenkō* (seniority promotion) system. *Shikaku Seidō* and *Nōryokushugi Kanri* were useful to some extent in solving the problems of rising labour costs and decreasing incentive due to the negative effects of seniority practice in long-term employment. They were also effective in dealing with the shortage of posts in relation to the number of people qualified for such posts. The termination of growth has two implications for management. One is the absolute shortage of posts and the other is the problem of labour costs. The basic idea of *nenkō* promotion, which is a collective promotion of employees with the passage of years, does not suit conditions of low economic growth, and a change to 'individualism' and 'specialisation' is a condition for flexible and individual personnel management. The issue is how to translate this idea into institutional reforms. In order to deal with the shortage of posts, the relative importance of individual assessment was increased, and in order to deal with the automatic increase of labour costs the percentage of seniority-based payment was decreased. All the major steel companies have adopted this method for both blue- and white-collar employees as well as for managers.

When a corporate structure begins to change from single-production oriented organisation to that of multi-profit oriented organisation, its management concept must also change to an individual and specialist-directed approach. In personnel management they must diversify employment practices and working conditions. The small number of core employees who are promised promotion to line posts will remain, while other employees may leave the company or be transferred to other organisations and then new rules and practices will be applied. Even among those who remain there will be a variety of employment conditions based upon individual assessment rather than age and length of service. Even employees of the same company may be getting different salaries and working under different conditions depending upon the business division they are in. This represents a large step away from the 'traditional' concept of life-time and collective employment towards a new concept based upon individualism and varying conditions of employment. This seems a sign of convergence towards Britain in the area of more profound principles of business organisation.

The role of foreman

In previous chapters we have seen that the roles of the foreman in Japan and Britain are quite different and there does not appear to be any convergence between them. The Japanese foreman plays a crucial role in linking the workers and management, rendering them key figures in the creation of harmonious industrial relations. It must be asked whether recent changes in the corporate environment have affected the role and function of the Japanese foreman, and indeed, there do seem to be some pressures for change in this area; in theory, this would seem to be another sign of convergence to Britain.

Pressure comes from two directions. One is from technological requirements and the other is from changes in the corporate environment, mainly from socio-cultural influence. Japanese foremen used to have two qualities, one an experience-based knowledge deriving from day-to-day work at the workplace, and the other a technological knowledge obtained through in-firm education and training. With the development of automation, a large part of experience-based knowledge and skills came to be replaced by computer programmes, and operational jobs have been standardised and simplified, as was examined in Chapters 6 and 7. This has, however, created a gap between the knowledge of sophisticated equipment and the operation of the equipment itself, which has become simplified and standardised. Young workers who would become future foremen feel it increasingly difficult to learn and be responsible for what is behind the computer panel board, i.e. an advanced knowledge of automated equipment. In terms of the cost of education and training too it is becoming more economical to make a division of labour between operators and engineers and leave the more technical aspects to the latter.

Socio-cultural influence, the feedback of industrialisation itself, should also be considered, as the steel industry is now in a state of maturity. When it was a rising industry in the national economy, it could recruit the best of high school graduates; but with the shift in the industrial cycle it can no longer recruit such a high quality of school-leaver. Moreover, as the percentage of those who go to university increases the students likely to enter the industry from school would tend to be lower in educational standards than previously. It is more realistic to adjust the use of the workforce to the changing aspects of corporate environment, namely by division of work. Foremen do not need to be high school graduates and recent ideas have shifted towards the foreman being a man from the workplace. Indeed, foremen recruited in the early 1960s will have already reached the age of retirement. To compound matters, university graduate employees are no longer interested in doing work on the shopfloor to obtain practical knowledge of day-to-day operations. The line and staff systems introduced in the late 1950s, in which foremen used to play a crucial role, will have to be seriously reconsidered. Management, and in particular those in human resource management, wishes to maintain the current foreman system and enhance the post by sophisticated education and training. This does not, however, seem appropriate when general education and training itself is beginning to change, along with the modification of long-term employment. One manager of a Japanese steel company told us that in future a university graduate would take on the role of foreman for some period in order to learn about the workplace and operational experience as part of his career.[7] Is this not similar to the experience of the BSC? It therefore seems possible that some convergence of Japan towards Britain may occur in this field as well, when the Japanese steel companies begin to adopt the more drastic corporate strategies required at the stage of contrived compatibility.

RESTRUCTURING IN BRITAIN[8]

Privatisation

The major form of restructuring in the BSC was privatisation, which meant a return to private ownership with a strong managerial initiative for efficiency and profitability. The privatisation of the BSC was due to a political decision of the Thatcher government to reverse the 1967

Labour Government Act. Each time such a big decision was made, the government presented a plausible explanation why such a change was necessary and what the implications of such change would be for the national economy. The privatisation of the BSC was one of a series of privatisation measures imposed upon British industry[9] which took place in the 1980s under so-called Thatcherism, whose other features included adopting economic policies of deregulation, the pre-eminence of the market economy, and tighter public spending. The key concept of such policies was monetarism and popular capitalism. Therefore the necessity for privatisation was explained in two ways; one, that spending public money for an ailing industry is not good, and the other, that efficiency would increase under private ownership by virtue of the competitive pressures of the market. Competition among rival companies and the pressures from customers and shareholders are considered to be an effective means of creating an efficient organisation.[10] The following constraints which were thought to have existed for the BSC prior to denationalising the steel industry in December 1988 (BS plc: Interview) were intended to be alleviated by privatisation:

- limitation of business operations within the domestic market;
- limitation of production to mass-production products;
- hindrance of business diversification;
- disincentive to make customer contacts, and
- discouragement of high-value-added products.

Corporate restructuring

The corporate restructuring of the privatised BSC, which thus became British Steel plc, can be examined in terms of two areas: one is that caused by corporate strategy and the other the restructuring required for managerial strategy.

The major corporate strategy included the acquisition of the largest stock company, C. Walker & Sons (Holding) Ltd, which resulted in the new private company controlling 34 per cent of the domestic stock market.[11] This was done to avoid the possibility of take-over by some other European steel company and also to expand the redistribution network in Europe. Another business strategy relevant to their production was the merger of their stainless steel business with Harvester of Sweden.[12] This was thought suitable as it would help to reduce costs and increase competitiveness. Another aspect of business strategy was to carry out restructuring in a number of works, such as Corby and Ebbw Vale. Shotton was re-organised as a finish process factory and

Clydesdale (Scotland) ceased production of pipes. It was decided to import pipes from Japan and process them in Clydesdale into finished products. Ravenscraig, one of the five major works, was closed in April 1991, resulting in 700 being made unemployed. This closure was explained by Ravenscraig's relatively inferior conditions in location and efficiency.[13] Other strategies were to establish separate companies within the precincts of steelworks and control them indirectly through share holding. Such concerns are Allied Steel Ltd, Guest Keen and Nettleford Ltd and United Merchant Bar plc, all manufacturers of rods and bars. British Steel plc owns 60 per cent, 40 per cent and 50 per cent of their shares respectively.

British Steel plc re-organised itself into four divisions: strip products, general products, distribution, and R & D businesses. In the case of the strip products division, Ravenscraig, Port Talbot, and Llanwern works are business units, while other smaller works or factories are considered simply as cost centres. In general, products divisions such as Scunthorpe and Teesside are businesses, while Corby and others are cost centres. The distribution division is a business which controls both domestic and overseas distribution.

The above business strategy shows a similarity to Japanese steel companies. The style of rationalisation, such as closure and reduction of employees, is similar, and older and smaller works come first. These works and factories have relatively inferior conditions, vis-à-vis both location and facilities, compared with modern large-scale works. At the Kamaishi works of Nippon Steel Corporation, equivalent to the British small, old works, a new joint company has also been set up.[14] All these rationalisation measures adopted in both Britain and Japan suggest that a logic of capitalist production, and in this sense the direction of convergence, is mutual. In addition, the system of division and businesses, namely the type of corporate structure adopted by the BSC and succeeded to by British Steel plc, clearly shows that Britain is far advanced in its form of corporate structure. The prototypes of the current division and businesses appeared as early as 1967 at nationalisation, and since then have undergone various reforms. The form of business organisation in the 1990s is another reformulation of corporate structure, made on the basis of contrived compatibility in management in response to the changing corporate environments.

Expansion of sub-contracting

After privatisation the management succeeded in further expanding sub-contracting, both in scope and depth. Although it had initially been limited to peripheral areas that either did not fall under the BSC remit or were unprofitable for BSC direct action, following denationalisation sub-contracting made inroads into the main process areas where regular employees of British Steel plc were and remain responsible. Some high-wage jobs, such as packing and scarfing, were also sub-contracted. In one case, regular workers attempted to prevent sub-contracting by proposing their own work rationalisation plan, but the management took advantage of that proposal to carry out a larger-scale rationalisation (Interview 1992). It also happened that those who became redundant on Friday were immediately re-employed as sub-contract workers or as groups of self-employed individuals. On another occasion management obtained quotations for certain jobs and presented their workers with the ultimatum to reduce their rate to an equivalent level or leave. This type of rationalisation is a clear indication that workplace organisation (work practices, manning, grouping) is a result of a balance of power between management and labour. A prime objective of managerial strategy is to reduce labour costs. The same workers may perform broadly similar jobs – but as employees of a sub-contracting company they must do it under inferior working conditions such as no pension scheme, longer working hours, insufficient manning and no compensation for injury. This is very similar to the fate of Japanese workers who were transferred to sub-contract companies to do similar jobs in the same steelworks.[15] This kind of degradation of working conditions was taken to the European Court and British Steel plc was ordered to alter it (Interview 1993). This alteration was included in the Trade Union Reform and Employment Rights Bill. Despite the fact that the sub-contract workers who work in British Steel plc are union members, they are not represented at the workshop. This makes it easier for sub-contract companies to reduce their working conditions. The value of sub-contracting to management is especially high in the following areas: reduction of labour costs; flexible use of labour; rationalisation of administrative work;[16] and increased control of remaining staff.[17]

The foreman system

The second aspect of restructuring was the rationalisation of white-collar employees, namely staffing reductions, which increased from 27.2 to 33.7 per cent between 1979 and 1992. This was carried out by the abolition of supervisory posts, in particular those of foremen. The major function of foremen was to oversee operations; the system had been introduced in the 1960s, but now began to be seen as unnecessary and costly. This implied a return to former working groups headed by work leaders such as first ladelman, first hand teemer, first vesselman, and first mortar. The major objective of the abolition is to reduce labour cost and to create direct and smooth communications between workers and middle management. Foremen are apparently viewed as if they have been blocking the channel of communications. In Britain this may well be true, but in Japan it was the foremen who became communication bridges between management and the shopfloor. It is thought that the elimination of foremen will improve communications and enhance administrative efficiency.

Thus we can see more rationalisation taking place at the lower levels of the managerial hierarchy in the 1990s. The case of Japanese foremen, discussed above, suggests that the role of foreman cannot be thought constant, and will change in response to the forces of corporate environments. Britain may be relatively advanced in this respect, and Japan may well converge towards Britain at some time in the future.

Local Lump Sum Bonus

The Local Lump Sum Bonus Agreement is similar to the Japanese annual bonus scheme in that the amount of bonus is linked to company performance. However, this lump sum bonus is more directly linked to the performance of the works in which the employee is based (Scunthorpe Works, Interview 1992). This scheme was introduced in 1980, when a three-month strike ended with a defensive compromise of the unions with management. It aimed to reduce the importance of annual wage negotiation as well as national negotiation by linking quarterly bonuses and wage increases to the objectives of rationalisation, such as production level, manning and flexibility. The maximum local lump sum bonus is determined at 18 per cent of the total payment. At first, the resources for this bonus were derived from workforce reduction, but now differ from establishment to establishment, and are variably dependent upon productivity, profits, and the amount of production. In practice, an enabling agreement is made and with this as a basis the percentage is fixed at a single-table negotiation. For example, if a worker's wage is £300 a week it adds up £3,900 in a quarter of thirteen weeks. Added to this is a £700 LSB at a maximum 18 per cent payment. In total it becomes £4,602, i.e. £352 a week. In the early years of this scheme, the majority of employees were getting the maximum rate of 18 per cent, but in recent years this has reduced. The reason for this is that part of the LSB has been consolidated into the basic wage. For example, the wage rise in 1992 was 2.5 per cent plus 1.5 per cent of LSB. In other words 1.5 per cent was transferred from 18 per cent of

LSB, resulting in a maximum LSB of 16.5 per cent. In order to get it back to 18 per cent, workers have to accept various kinds of changes in work practices such as in manning, team working, overtime work, etc. (ISTC Interview 1992). This condition is stipulated in the regulation of payment which is a part of the labour agreement. In negotiation the management will investigate the relations between the changes in work practices and the rise in productivity. As workers have taken this bonus as part of their regular pay, they must concede further changes in work practices or the maximum level of LSB will get smaller and smaller. This is a sophisticated method of labour control by means of wage leverage, which demands self-regulated rationalisation as an incentive for workers who are lucky to have escaped redundancy.

Thus, the LSB is a typical form of productivity-linked incentive payment which appeals to workers more directly than the Japanese bonus system. The idea of the bonus itself seems to suggest a convergence towards Japan, but the actual method may be introduced into Japan in the future when the corporate structure assumes a full-scale division and profit centre structure along with the change in the nature of industrial relations.

Labour flexibility

Work practices have also changed in the 1990s and the major trend has been to create flexible working practices. Work and employment practices are interrelated, forming an important area of management. Work practices are also relevant to working conditions and this relationship is regulated by the nature of industrial relations (see Figure 1.1). In this sense, any change in work practices is an indirect consequence of industrial relations, in particular those active at works level. The flexible use of the workforce in terms of a combination of various workforces such as sub-contracting, part-timers and other casual workers is one kind of flexibility, brought about by change in employment practices, while flexibility among regular employees, such as group working, is a direct consequence of changing work practices. Here we examine flexibility among regular employees and its connection to industrial relations, and consider its implications for convergence.

Managerial strategy for flexibility was to propose the following changes in work practices to the union (Scunthorpe Works, BSC 1992):

1 flexible assignment of working days within the same shift group;

2 training of operatives and craftsmen for both crane operation and maintenance;
3 training of operatives and craftsmen for regular maintenance;
4 promotion of QC circles;
5 agreement on the method of operation, performance and manning on the introduction of new equipment;
6 transfer of older workers to simple jobs with lower wages, while maintaining current wage levels until other wages rise.

The agreement with the union was to create flexibility in those areas where only one union had influence. In future, further flexibility will be pursued in areas where more than one union is involved. This kind of flexibility in labour was made possible due to the effect of technological innovations. One effect was the standardisation of job content, which deprived workers of job control, which was transferred to both machine and management.[18] The other effect was the absolute as well as relative decrease of manual workers who were the core of traditional union power and this provided conditions for tilting the balance of power for flexibility.

Industrial relations[19]

The nature of industrial relations in the British steel industry from nationalisation in 1967 up to the late 1970s can be characterised as 'co-operative' on the basis of industrial democracy, namely the co-existence of capital and labour. In this period, the number of employee directors increased, reaching a peak in 1979, as noted in Chapter 10. However, this relationship changed towards the end of the 1970s and resulted in the 1980 strike. Management used this strike and the political climate in Britain at the time as an opportunity to seize the power initiative, although relations at the everyday level continued to be conflictual until about 1990. During the 1980s a number of legislative measures enacted by the Thatcher government strengthened the legal framework. This weakened the trades unions and the strong union influence of the 1970s waned, resulting in the revision and degradation of what is stipulated in national level agreements. In the 1990s, a more 'compliant' nature of industrial relations has emerged under the increased authority of management, confirming the establishment of powerful managerial authority.

In the 1990s, the emphasis of national level negotiation has shifted to the harmonisation of working conditions between white- and blue-collar employees,[20] the fair distribution of the profits resulting from employee education and training, the provision of pension schemes and the acquisition of business information, etc. At establishment level, negotiations involved wages, working hours, holidays and sick leave. These adopted what is commonly called a single-table negotiation, in which all the relevant unions in British Steel plc come together at the same table and what is determined there is accepted by all the unions concerned. Thus, all the practical issues are settled at the Business Joint Council at each establishment. These deliberations are monitored by the Central Joint Co-ordinating and Monitoring Committee, which is held at regular intervals. This committee discusses problems relevant to the agreement as well as their solutions. In particular, issues relevant to more than one union become a major topic. Some of the problems beyond the range of the agreement are solved according to the way of negotiation that existed prior to this new method.

We can note three factors in the increasing management initiative in modern industrial relations.

The first is the already-mentioned legislative attack on the unions by the Conservative government. One example of the effect of this is that it now takes much longer to plan a legal strike, giving management ample time to prepare for it. Workers will be dismissed if they strike illegally and the union has to present evidence that it is not involved in it.

The second factor concerns the financial situation of the union movement. Any recession seriously affecting the health of the manufacturing industry brings about a reduction in union membership, which in turn affects the financial stability of the union. However, ISTC, although membership has shrunk considerably, has sufficient assets, which produce a relatively good return from interest. Thus, the reduction of union membership has not directly led to financial crisis in the main union.

The third factor is the attitudes of the union members. Since the strike of 1980 union members have lost their sense of resistance, in spite of their increasing anxiety and dissatisfaction, due to the pressure of high unemployment created by the long-running economic decline.

The above conditions which the union members have had to face seem to have contributed to increased managerial authority. The workers who remained in employment now have to accept more responsibilities than formerly, lower manning levels, more training and education, more frequent calls for emergency work, tight quality control and longer working hours due to the change from twenty-one to fifteen shifts, etc. Thus, a degradation of working conditions in various forms seems to prevail unless the unions, with the support of rank and file members, manage to regain the power to counterbalance managerial authority. This shift in the balance of power has tended to create a phenomenon of convergence towards Japan in the area of work practices and working conditions. Thus the case of British Steel plc after privatisation suggests that a similar trend in technological paradigm can create a convergence in work practices and working conditions if union power is weak and the balance of power shifts in favour of managerial interests.

CONCLUSION

The most recent changes in the 1990s in both Japan and the UK suggest that a common approach has emerged in managerial strategy for dealing with the similar macro-economic conditions faced, and this has resulted in further convergence, in particular from Japan towards Britain. The rapid changes implemented by Japanese management meant a reformulation of their managerial principles applied to a small head office, division system, business diversification and overseas business, which are related to the organisational restructuring of the company. The changes in employment practices and the role of foremen are relevant to the management of employees in terms of control and integration. The nature of such a reformulation suggests a shift from collective to individual control, and from an orientation towards productivity and growth to one based on subdivision of product lines and an emphasis on profitability. The shift began to emerge with the ending of high economic growth in the mid-1970s and continued thereafter, but became more explicit and distinct through the 1980s into the 1990s. This shift has also involved a move from organic to contrived compatibility in management and an effective termination of the circumstances conferring relative advance, which had been based upon an advanced production system.

Concurrently, in Britain the change has also been drastic but within traditional managerial principles. The British response to the macroeconomic influences of the 1990s can be identified in two areas. One area includes an expansion of sub-contracting and flexible use of the workforce. This has been made possible by new technological conditions and union compliance. The other is a consequence of their response to the change of ownership from public to private, and has appeared as organisational restructuring.

In both Japan and the UK such drastic restructuring was promoted by the enhanced power of management, which conversely implied the relative increase of union compliance. This situation in Britain was largely induced by political measures, but was more voluntarily assumed by Japanese unions. The new production system, together with other factors, undermined the traditional power of unions both in Japan and the UK. This was due to two factors: first, the decrease of that part of the workforce who were the core members of the unions, and second, the change in work content, which reduced the relative autonomy of the workers founded upon experienced-based skills. This allowed management to increase the integration of the workforce through various practices, rules and regulations. Replacement of older workers with younger ones and a readjustment towards the enhanced integration of workers have been common ingredients of management strategy in both industries.

The hypothesis underlying this work has been that, regardless of initial differences in corporate management and labour, the adoption of similar production systems will provide an impetus for convergence, and a shift of industry from growth to relative decline will also create conditions for convergence in corporate structure. Thus, while Japan enjoyed a state of relative advance from the 1960s, due to the new production systems introduced at that time, the delayed introduction of such systems into Britain in the latter half of the 1970s provided a stimulus for Britain to converge towards Japan. At the same time the termination of growth and the similarity of macro-economic conditions in Japan provided the momentum for subsequent Japanese convergence towards Britain in managerial orientation and corporate structure.

Thus the research presented here, although limited to two steel industries, points to twin forces creating a universal trend in management and labour; one representing technological conditions and the other macro-economic conditions. This is not to deny the existence of diversity in management and labour in both the national and international dimensions. Various differences will persist, as all the factors in Figure 1.2 will never completely converge either quantitatively or qualitatively. So long as technological and economic conditions differ in form and extent, we are unlikely to experience any final and complete convergence from either side in management and labour. Much depends on the decisions of human beings who, even in a similar corporate environment, will respond in individual and culturally influenced ways. This is so on both individual and collective levels, as found in the different response of employers' associations and labour unions. There is thus ample room for different dynamics of compatibility in management and labour. This world of diversity within convergence is a general conclusion identified in this research.

Today a large number of Japanese steel employees, both blue- and white-collar and including managerial staff, who were once devoted to corporate objectives and ideology, have left or will leave their companies with mixed feelings of betrayal, resignation and disappointment. But these employees are beginning to develop a stronger sense of individualism and a sense of pluralism from their experience of sacrifice to their companies. Similarly, those who still remain in management regard these values as necessary in order to control and integrate employees in times of low growth and profit orientation. The era of Japanese-style unitarism in industrial relations, for both employees and company, which supported the organic compatibility of management in the Japanese steel industry, is gradually being replaced by a tentative social trend towards a version of individualism and pluralism. Concurrently, in the British steel industry, located far down the road of contrived compatibility, traditional practices are being replaced gradually in many workplaces with union compliance. These are often considered similar to the 'Japanese practices', albeit with no apparent change in the traditional perspective of pluralism in industrial relations.

So, where might the iron and steel industries of Japan and Britain go from here? Given the current state of the world economy and the general downturn in heavy manufacturing, as well as the competition of cheaper labour markets, steel enterprises of the more mature economies seem to have no choice but to trim their sails to the wind, continually adjusting their systems of management and corporate structure to keep abreast of the corporate environment. But when we view the steel industry in the long term not just as an economic entity but also as an important social unit of production and employment, short-term adjustments aimed at immediate profitability may not be sufficient. The management and labour of such an important industry as steel, when it is at the stage of contrived compatibility, must be considered more in the context of long-term social benefits without wasting either physical or human resources. In any case, the prospect for both national industries is one of strenuous effort to maintain a reasonable level of continued compatibility in management just to fulfil the primary requirement of profitability.

Is any alternative scenario reviving conditions of organic compatibility possible? One may speculate on the possibilities if a new industrial philosophy, which might be called 'socio-economic democracy', were to emerge as a response to the changing corporate environments. To be effective, such a philosophy would have to accommodate the principles of efficient production, a democratic balance of authority and reward, and reasonable working conditions for the workers, which would need to be in proportion to increased productivity as consensually agreed among management, workers and unions. Accommodation must also be made for environmental protection measures which are becoming increasingly required in the modern industrial climate. At this time, however, such a shift in perspective has yet to emerge. If, therefore, management and labour in the Japanese and British steel industries are to enjoy a future offering more than a mere struggle to maintain contrived compatibility, both must begin to ask searching questions. The present writer can see no other way to find non-traditional answers to the goal of ensuring continuing viability for the steel industries in the twenty-first century. At present it would be asking a great deal of organised workers to expect them to participate and co-operate fully in a process which inevitably involves substantial job losses of which they will bear the brunt. Of course, this would be made much easier if government were able to improve the macro-economic environment and the possibilities for retraining so that redundancy were no longer so frightening.

At the current point in history, the Japanese and British iron and steel industries have one major thing in common – in order to thrive, they need to apply strategies that are both positive and innovative in both economic and social requirements.

Notes

1 INTRODUCTION

1 In Japan during the period of this study the development of the steel industry was a precondition for the development of other heavy industries such as shipbuilding, machinery and automobile production, while in Britain, although the major steel-consuming industries were the automobile and machine industries, the decline of heavy industry as a whole seems to have distracted attention from comparison with Japan. Apart from this kind of comparative study there exist some important studies such as those by Ovenden (1978), Bryer *et al.* (1982), Hartley *et al.* (1983), Abromeit (1986).

2 In the period 1989–93, the total number of overseas investments by the five major steel companies was twenty-nine, of which twenty are in the United States in the form of joint ventures (Tekkō Renmei 1993). The most active companies are Kobe Steel Ltd and Sumitomo Metal Industries Ltd, while Nippon Steel Corporation undertook rather few of these ventures. Two representative cases of Nippon Steel are with Inland Steel Industries: one to produce cold-rolled steel products (I/N Tek) and the other to produce hot-dip galvanised and electro-galvanized steel products (I/K Kote) (Nippon Steel 1992: 8). This joint venture aims to supply Toyota, Nissan and other US car makers. Two reasons exist as factors hindering overseas investment by the steel industry: one is the amount of capital and technology such a large and sophisticated industry requires, and the other is that the steel industry is generally regarded as a nationally important industry. For these reasons, most overseas investment by the steel industries is limited to small- and medium-sized investment for rolling processes.

3 A worker criticised the situation as follows: 'At the workplaces where we worked many senior workers are now being phased out . . . . Is it something we should allow? Although the enterprise as it is is even now growing' (Rōdō Tsushin 1987).

4 It is interesting to note that in 1993, when major steel corporations began their drastic rationalisation to cope with environmental changes, Mr T. Imai, the president of Nippon Steel Corporation, admitted the decline of the steel industry in reference to his own company: 'I believe that the Nippon Steel Corporation, as an enterprise, passed its peak as many as twenty years ago' (Nippon Steel, *SHINNITETSU* 1993, Vol. 76: 11).

5 Hunter *et al.* (1970) investigated the impact of changes in technology upon

employment in three selected industries: printing, steel and chemicals. The work of Forslin *et al.* (1979) was an international comparison of the effects of automation on production workers; attitudes of workers, work content, and working conditions. Ozaki *et al.* (1992) focused upon the international comparison of the effects of micro-electronics technology upon management–labour interaction in three industries: machinery manufacturing, printing and banking. Their common interest and understanding was that technology affects labour and management and this is an important issue both in theoretical and practical implications. This study, which shares this perspective, attempts to further develop the area of the above studies with its own theoretical assumption in the Japanese and British steel industries.

6 In this connection, Dore (1973) investigated the origins of national diversity in industrial relations by the use of the convergence thesis and that of a late development effect. While Dore emphasises that Japan is a case of late development in the area of industrial relations, this study uses the convergence thesis to explain the relative advance of the production system and corporate structure.

7 The convergence thesis used in this period can be called the 'industrialisation paradigm' of the 1960s and Dahrendorf (1959), Rostow (1960) and Bell (1960) may be considered other examples (Inagami: 1984).

8 One of the important areas in the convergence thesis adopted by industrialism is the demise of class conflicts and development of corporatism between employers, labour and government.

9 The focus of criticism was on Braverman's view of the working class as a class 'in itself' rather than 'for itself' and his pessimistic determinism regarding the technological effects upon the working class. The dynamic and dialectical aspects of social relations are the points emphasised by critical views such as Tomizawa (1978), Yamaguchi (1979) and Watanabe (1978).

10 Technological condition is defined rather broadly, for the purpose of this study, as a production system shown in Figure 1.1. It comprises production technology, work practices and working conditions which together form a manufacturing process and method. The nature of technological conditions we refer to in this study is a highly automated production system as represented by large, continuous, high-speed and large-scale steel production with the use of sophisticated computers.

11 The new technological circumstance is defined rather broadly as the higher level of automated production system in which the production system itself has some relative advantage over others. It is usually accompanied by compatible work practices and working conditions as shown in Figure 1.1. They are, however, constrained by the process and nature of employment practices and industrial relations, rendering complete convergence unlikely even under similar technological circumstances.

12 The British Steel Corporation sent senior management to conduct a thorough study of the management system in one of the major steel companies in Japan in 1983. This study seems to have been used to reconsider their management system under the new technological condition.

13 As shown in the chapters in Part Two, the case of Japan typically illustrates how macro-economic conditions were the primary factor for corporate growth and the established institutional arrangements, often conceived as

a cause for corporate growth, were the consequence of a long-term stable corporate growth. But the fact is that, once established, they tend to hinder the efforts of managerial response to the changing corporate environments.

14 An early study on this issue is found in Bright (1958), in which he states that 'The economic evaluation of automation, therefore, lies in determining all the benefits and all the liabilities that are likely to accrue to the particular automation project' (Bright 1958: 143).

15 The function of in-firm education and training as an instrument of control and integration is often overlooked and emphasis is placed on the technical utility of the workforce developed by such programmes. For example, whether Japanese in-firm education and training, in particular that of white-collar employees, has any relative superiority in international comparison or not has been recognised as an important issue in the 1990s (Japan Institute of Labour 1993: 1–73).

16 As will be examined in Chapter 6, these are the workers working on the same site side by side with the employees of the steel companies. This is due to the perspective of study which defines the scope of management in a narrow sense without paying attention to the significant role of sub-contracting in terms of the social division of labour as well as the practical benefits deriving from it (Ishida 1981: 2).

17 The development of the steel industry in the UK was consequent on domestic technological inventions such as the Bessemer Converter in 1856 and also the expansion of markets due to the diffusion of the industrial revolution to other parts of the world.

18 In this period, characterised as the Victorian Age (1837–1901), liberalism and competition were thought to be the driving forces and values for growth and development. Trevelyan characterises conditions for such an age as 'no fear of catastrophe from without' and 'seriousness of thought and self-discipline of character, an outcome of the Puritan ethos' (Trevelyan 1978: 448). As regards their business attitude, we might regard it as more production-oriented than now and successive investment created the stimulus for day-to-day inventions and innovation which led to the increased competitiveness of their products. Britain was indeed a role model for industrialisation across the world.

19 In the case of Japan they acquired raw materials, technology, management techniques and some finance from abroad; in particular, the USA played a crucial part, as is investigated in Chapter 5.

2 ECONOMIC DEVELOPMENT

1 Data concerning the figures used in this section derive from the following publications: IISI 1972; ILO 1992; Nihon Ginkō Chōsa Tōkeikyoku 1970, 1981; Keizai Kikakuchō 1980; Tekkō Renmei 1969; UK Iron and Steel Statistics Bureau 1960, 1967, 1970, 1975, 1979; Tekkō Renmei (Tekkō Tōkei Yoran) 1970, 1975, 1980.

2 According to the ILO report the steel intensity of Japan went down sharply in the period 1970–80, suggesting a correspondence with the two stages of system compatibility of management shown in Figure 1.3.

3 The levels of GNP and steel production are not always correlated. For

example, Switzerland and Canada and some other countries do not have a large production of steel in spite of their high level of GNP (Nihon Ginkō 1986; Tekkō Renmei, Tōkei Yōran 1990).

4 The data for figures in this section derive from the following publications: Tekkō Rōren, Rōdō Handbook 1981: ILO 1966; Rōdōshō, Rōdō Hakusho 1960, 1965, 1970, 1975, 1979: Department of Employment 1971; CSO 1973, 1980, 1981; UK Iron and Steel Statistics Bureau 1960, 1981; Tekkō Shimbun-sha, Tekkō Nenkan 1970, 1975, 1980; BSC 1967/68, 1972/73, 1976/77, 1978/79, 1979/80, 1980/81.

5 The data for figures for this section derive from the following publications: Tekkō Renmei, Tōkei Yōran 1970, 1971, 1980; Nihon Ginkō 1981; Tekkō Shimbun-sha, Tekkō Nenkan 1962; UK Iron and Steel Statistics Bureau, Annual Statistics 1960, 1967, 1970, 1975, 1979; BSC, Annual Report and Accounts 1976/77, 1978/79, 1980/81; SCNI (Select Committee on Nationalised Industries) 1977a.

6 The percentages for 1973, 1974 and 1975 were 14.4 per cent, 19.4 per cent and 18.2 per cent respectively, constituting the top export item in terms of monetary value (Nihon Ginkō 1982).

3 INDUSTRIAL ORGANISATION

1 The degree of institutional shareholding of the steel companies is almost the same but they all belong to a different financial group and they are one of the core enterprises of each group. Nippon Steel Corporation has equal distance with major banks, making itself a huge industrial capital group (Okishio and Ishida 1981).

2 For example, Mr Y. Inayama who graduated from Tokyo University in 1927 entered the Yawata Seitetsusho (Yawata Steel) in the same year. He became managing director in 1950, vice-president in 1960, president in 1962 and when Nippon Steel Corporation was formed in 1970 through the merger of Yawata and Fuji Iron and Steel Corporations he became its president and in 1973 took the position of chairman at the age of 69 (Nippon Steel, Yukashōken Hōkokusho 1977).

3 The initial idea for this merger emerged around the time when the recommendation on the future steel industry was issued from Sangyo Kozo Shingikai (Industrial Structural Council) of MITI in 1966 (Ichikawa 1969: 193). The Fair Trade Commission issued a recommendation in 1969 that the merger contract between the two companies should not be exercised. Academics were also against the merger on the ground that it will restrict competition and inhibit efficiency, and will create an unfavourable condition for workers through rationalisation and monopoly prices (Mainichi Shimbun-sha 1969).

4 The nature of oligopolistic competition in the context of the Japanese steel industry at first resulted in the increased share of late developers (Kawasaki, Sumitomo and Kobe), and this threat in turn encouraged the capital investment of the early developers (Yawata, Fuji and Nippon Kōkan).

5 This rise continued through the 1980s due to the lower cost of scrap, allowing the electric furnace enterprises consistently to account for just under 30 per cent of steel production (Tanabe 1981: 216–22).

6 Kawasaki Steel began the operation of Chiba works in June, 1953; Sumitomo, July, 1953; Kobe Steel Dec. 1954 (Tekkō Renmei 1959: 123).
7 The imports/exports ratio in the period 1971–3 was 141.1 per cent, while in 1974–6 it jumped to 331.1 per cent (UK Iron and Steel Statistics Bureau 1970; 1975; 1979).
8 The market share of BSC in finished steel declined from 68.6 per cent in 1967–9 to 48.0 per cent in 1979.

4 INSTITUTIONAL PRACTICES

1 How management and labour may change in response to the current stage of industrialisation will be examined below. It will be shown that the system known as Japanese 'life-time employment' – comprising such features as long-term employment, seniority promotion, seniority-based wage/salary increases, corporate welfare, collectivism and enterprise unions – was highly functional and logical in the large-scale enterprises during the rising stages of industrialisation, but was essentially a consequence of that industrialisation rather than its cause. The termination of growth heralded decline in these features. Although antecedents may be traced in the pre-war period, this writer feels it is adequate and more relevant to see the post-war emergence and decline of the above features as arising from post-war conditions.
2 In this connection, recruitment, transfer, in-firm education and training, personnel assessment, job analysis, dispatch, retirement and re-employment, which would all be included in the category of employment management, become more important than they are in British management.
3 According to an interview with a union official the 'needed personnel' quota for skilled workers was determined by the union.
4 The legitimacy of sub-contracting lies in the benefit of sub-contracting itself, in particular at the stage of 'positive' compatibility. It is a division of labour based upon the logic of the parent company and to that extent the freedom of sub-contracting firms is limited.
5 Among sub-contracting firms there are relatively large companies which operate in more than one steel works and small local companies which operate only in one works. Higher dependence in the amount of work creates stronger affiliation with the 'parent' steel companies (see Chapter 5 and also Michimata 1978).
6 Data from the BSC (1986).
7 Interview at the ISTC (1986).
8 Data from the BSC (1986).
9 Data from the BSC (1986).
10 Data from the BSC (1986).
11 Data from the BSC (1986).
12 Data from the BSC (1986).
13 This is a package of economic programmes carried out by the occupation forces in Japan in January 1949, which included nine economic principles, the fixed exchange rate of one dollar = 360 yen, measures to curb inflation and the tax system.
14 Each level of union hierarchy matches that of management in terms of

consultation and negotiation: *Shibu* (branch) with the head of section in the works, *Tanso Honbu* (headquarters of works) with the head of works, *Kiren* (federation of works) with the head office of the company and *Tekkō Rōren* (Japan Federation of Iron and Steel Workers' Unions) with *Tekkō Renmei* (Japan Iron and Steel Federation) .

15 First Report from the Select Committee on Nationalised Industries, Session 1977–8, British Steel Corporation, vol. II, HMSO, 1977a: 20.
16 Interview at the Scunthorpe works (1986).
17 Interview at the Scunthorpe works (1986).
18 Interview at the Scunthorpe works (1986).

5 THE JAPANESE STEEL INDUSTRY

1 There were special measures for promoting capital investment: a special depreciation measure for rationalisation equipment, a tax-exempt measure for vital imported machinery, a special tax measure for scrap imports, and government subsidies for research and development (Tekkō Renmei 1969: Chapters 5 and 7).
2 Yawata Works (Nippon Steel), Yawata Works Brochure 1977; Nippon Kokan KK 1972: 294; Kobe Steel Ltd 1974: 50–1; Wakayama Works (Sumitomo Metal Industries Ltd) Wakayama Works Brochure 1977.
3 Class A (*Kō-shu*) includes long-term technology imports such as the first technology import after the war of plate manufacturing from Armco Steel by Fuji in 1951 and by Yawata in 1952, while class B (*Otsu-shu*) includes purchase of blueprints and invitations of technologists for instructions.
4 The total number of technology exports was one in 1960, nineteen in 1965, sixty-nine in 1970, and 150 in 1975. The rapid increase of technology export since 1965 has become a strong means of promoting joint business, and securing raw materials abroad. The Japanese steel industry has exported its technology to as many as twenty countries (Tekkō Renmei 'Accommodation of Japanese Iron and Steel Industry towards Overseas Integrated Steel Works Projects' in *Tekkō Kaihō*, May 1976.
5 Figures are from the Report on the Survey of Research and Development in Japan (1965, 1966, 1969, 1975 and 1976) by the Sōrifu Tōkeikyoku and also Tekkō Shimbun-sha 1969, 1970 and 1976.
6 Nippon Steel Corporation Kimitsu Works provides useful information on their computer control system in *Kimitsu Seitetsusho Sōgō Jyōhōshori System* (Total Information System of Kimitsu Works).

6 MANAGEMENT AND LABOUR IN JAPANESE STEEL WORKS

1 Japanese labour productivity in steel production in 1950 was one-third that of the UK and one-seventh that of the United States (Tekkō Renmei 1959: 208). In Japan Nihon *Seisansei Honbu* (Japan Productivity Centre) was established in 1955 and the steel industry dispatched a team to the US in 1955 and 1957. The 1955 mission was to make a thorough survey of corporate organisation, sales, purchase, managerial accounting, labour and

production in the major US steel companies. They identified four reasons for the success of the US steel industry: abundant raw materials, large markets, concentration of enterprises, and modernisation of plant and equipment. They also learned about various management methods such as line and staff systems, industrial engineering, the controller system and marketing. The 1957 mission concentrated on learning technical aspects of steel production and its management. The knowledge of management brought back in this period provided the foundation for later management development in the Japanese steel industry, together with the large number of technology imports from the US. See Tekkō Renmei 1959: 143,208–12.

2 But this does not mean that the absolute importance shifted from blue- to white-collar employees. Emphasis on the successful management of blue-collar employees has been strong even as the content and nature of the work itself was undergoing change. The reason for this is that the increased continuity and speed of production processes required a nearly perfect integration and control of the workforce.

3 Although this department is quite large in percentage, its status is lower than production process departments and other technology-related departments. It also has comparatively low requirements in the areas of team working, creativity and technical knowledge.

4 The figures used in the above section are from Yawata Steel Corporation 1961 and Nippon Steel Corporation, Kimitsu Works 1975.

5 The control span in 1992 was 60 : 1, the number of managerial positions six, the percentage of *Buchō* (general managers) 18.6 per cent, the percentage of senior management 18.7 per cent, the span of control of senior management 5 : 3. In 1992 the total number of managerial personnel above the section chief known as *Shitsuchō* (room chief) in 1992 was ninety-one, the lowest number since 1960. The total number of employees was 5,477, a decrease of 261 from that of 1977 (Interview: 1993).

6 In 1992 we see a relative increase of senior management, which indicated that one in 3.8 of middle management has been promoted to senior management positions (Interview: 1993).

7 Kimitsu held the highest rate of sub-contracting of 66.3 per cent even in 1991, while Yawata was 58.4 per cent in the same year, which was a large increase compared with 44.4 per cent in 1977 (Interview: 1993).

8 One reason is that the workers in new works are younger than those in old works, as shown in Table 6.10. Another reason is that Kimitsu is nearer to Tokyo, where people have more individualistic values which have influenced those of surrounding districts (Interview: 1992).

9 Article 64 in the Japanese Labour Standard Law in principle prohibits employers from using female employees from 10:00 pm to 5:00 am.

10 These companies are called *Odan Kigyō* (horizontal firms), in contrast to *Jiba Kigyo* (local firms). *Odan Kigyō* is a firm which has establishments in more than one steel works, while *Jiba Kigyō* have just one local establishment. Nippon Steel Corporation had a total of thirty *Odan Kigyō* and 135 *Jiba Kigyō* in 1992 (Interview: 1992).

11 For example, the maintenance of drainage in the blast furnace, the changing of rolls in the rolling mills, and major maintenance work after an accident are typical sub-contract jobs.

12 A typical job is to check if actual figures on the panel board are within the

standard range and if any irregularity is observed to make a fine adjustment or new input in order to bring the figure back within the normal range. The experience and skills needed for operation, such as blowing and speed of rolling, are all computed into the standard models of operation (Interview: 1992)

13 This is a personnel management tool which is essential in *Nōryokushugi Kanri*. While *Nōryokushugi Kanri* is a broad framework of management which reflects the idea of human resource management, *Shokunōshikaku Seidō* is a more specific method of grading employees according to the in-firm qualification system.

14 *Jishu Kanri Katsudō* and other small circle activities began to spread with the formation of the committee on *Jishu Kanri* activities in the Japan Iron and Steel Federation. By 1983, forty-five out of forty-eight member companies of the Federation had taken part in *Jishu Kanri* activities and organized 29,150 groups of 183,648 workers (Tekkō Renmei 1985). These so-called 'autonomous' management activities, set up by management and with almost 100 per cent participation of production workers, imply several things in the context of management. First, the system suits group working established under an advanced production system that devalues the traditional skills and experience of workers; second, when labour unions are 'compliant', as in Japan, it is easier to introduce, and when labour relations are 'conflictual', useful to alleviate the power of unions; third, it is useful to mobilise workers towards the objectives of the works and the company through an enhanced mentality of co-operation among team members and that of competition among other teams. It is reported that of 2,077 subjects adopted in the QC circles of the Kakogawa works of Kobe Steel in 1990, 42 per cent contributed to safety improvement, 27.7 per cent to productivity increase, 10.6 per cent to equipment improvement, 9.1 per cent to quality improvement, 4.7 per cent to yield, 2.9 per cent to manpower and other improvements and 2.1 per cent to energy and resource savings (Kobe Steel, Interview: 1992). Workers' efforts in the 1970s and 1980s in this movement have resulted, ironically, in creating a relative surplus of labour, as the termination of growth has become more or less a constant condition for the steel industry. Drastic restructuring has taken place in 1994.

7 MANAGEMENT AND LABOUR AT HEAD OFFICE (JAPAN)

1 In terms of the organisational development of head office in the high-growth period (organic compatibility period) the steel enterprises can be counted as a typical case of a head office controlling an increasing number of production units (works) based upon a *similar* production technology. Head office serves as the administrative centre for non-diversified homogeneous line organisations. This functional aspect of the head office in the high-growth period created a strong centralisation, although after the growth period and in particular during the 1990s this style of head office began to be reconsidered to fit a more profit-oriented, diversified divisional organisation. For more general analysis see Kono 1984.

2 In 1992 the total number of top management of the Nippon Steel

Corporation was fifty-two in addition to three senior advisors who used to be influential members of top management. The breakdown of the fifty-two was one Representative Director and Chairman of the Board of Directors, one Representative Director and President, seven Representative Directors and Executive Vice Presidents, sixteen Managing Directors, twenty-three Directors and four auditors (Nippon Steel, *Basic Facts About Nippon Steel* 1992).

3 According to the 1994 restructuring plan for the whole organisation, this management policy council is to be merged with the managing directors' council becoming *Keiei Kaigi* (Corporate Managing Council) and executive functions are intended to be largely delegated to the line organisations (Tekkō Renmei, *Tekkō Kaiho* 1994: No.1618).

4 This dual function of Japanese top management is unique to Japanese corporations compared to the American and European corporations in which ownership, rule and control are more clearly identified in terms of function, people and organisation. See Nakata 1993; Scott 1986; The Economist 15 May 1994.

5 This development in 1978 began to be reconsidered in the Third Mid-term Corporate Plan (1994; 3) and its proposal is to merge the personnel and labour departments (Tekkō Renmei, *Tekkō Kaiho* 1994: No. 1618).

6 The same trend continued and we see the percentage of Senior Management increased to 18.6 per cent and Assistant to Senior Management to 48.5 per cent, which resulted in both a relative and an absolute drastic decline of middle management to 27.4 per cent in 1989 (calculated from Kaisha Shokuin Roku by Daiyamondo-sha 1979). However, this top-heavy structure of management is planned to be restructured by halving the number of head office employees (Tekkō Renmei, *Tekkō Kaiho* 1994: No. 1618)

7 This change in the content of office work and standardisation of managerial function at the middle layers of large organisations has contributed to an increased hierarchical management in which power is greatly concentrated in top management. Even decision-making at the every day level has tended to go through the bureaucratic process; first, planning at non-managerial level, then steadily upwards to *Kakarichō* (section chief), *Shitsuchō* (room head), *Buchō* (general manager), *Torishimariyaku* (director), *Jyōmu* (managing director), *Fukushachō* (vice president) and *Shachō* (president) before execution can go ahead. This too is being reconsidered in the 1990s (Nippon Steel, *SHINNITTETSU* 1994: Vol. 82).

8 Promotion below the *Buchō* (departmental manager) is based upon the 'plus points' principle while above *Buchō* it is based upon the 'minus points' principle. Apart from the formal process of promotion an informal and personal approach for successful promotion is important.

9 The aspects of participative organisation and self-actualisation in the human resource approach does not seem to be manifest in the Japanese in-firm education and training activities of the 1960s. The emphasis has been on the creation of employees to meet the needs of the enterprise at each different period. The philosophical aspect of human resource management appeared during the period of 'contrived' compatibility when growth almost terminated and *Nōryokushugi Kanri* was introduced. The British perspective for employees is more traditional in the interpretation of the relationship between person and work. See Ito 1994: 145–9, Iwade 1991: 158–63.

10 The *Shokunōshikaku Seido* divides employees first into broad categories such as blue- and white-collar employees and further into gradings. The maximum and minimum level of salary is determined for each grading. Also the average amount of increase is set. Each employee is assessed and ranked according to one of these gradings. The percentage of payment according to this factor began to increase in total remuneration from 12.4 per cent in 1967 to 60 per cent in 1988 (Hasegawa 1993).

8 THE BRITISH STEEL INDUSTRY

1 Data for figures in this section were obtained from the UK Iron and Steel Statistics Bureau. Data for figures for BSC in this section derive from the BSC Annual Report and Accounts, 1971/72, 1976/77, 1978/79 and 1980/81.

2 As of March 1980 there are the following subsidiaries of the Corporation located in the UK: holding companies (2), steel producing, further steel processing and allied activities (15), metal recovery and slag processing (9), building, construction and civil engineering (15), chemicals and chemical by-products (6), shipping and allied activities (1), companies registered under the industrial and provident societies acts, housing companies and statutory undertakings (9), other investments (10), export selling companies (3). As of March 1980 British Steel Corporation (International) Ltd or its subsidiaries, hold the following investments: England (8), Scotland (1), Argentina (2), Australia (1), Canada (7), Eire (1), Curacao (1), France (2), Germany (3), Holland (1), India (3), New Zealand (3), Nigeria (1), South Africa (6), Sweden (2), Switzerland (2), USA (1), Zimbabwe (2) (BSC, *Annual Report and Accounts*, 1979/80).

3 The Corporation has been in deficit since 1975 (1975, £255m; 1976, £95m; 1977, £443m; 1978, £309m; 1979, £545; 1980, £668m), while BSC (International) has been profitable with net profits before tax (1975, £23.2m; 1976, £19.2m; 1977, £12.4m; 1978, £14.2m; 1979, £19.8m; 1980, £22.0m) (BSC, *Annual Report and Accounts* 1980/81, 1981/82, BSC (International) Ltd, *Report and Accounts* 1979/80).

4 DTI, *British Steel Corporation: Ten Year Development Strategy*, 1973, Cmnd 5226.

5 DTI, *British Steel Corporation: the Road to Viability*, 1978, Cmnd 7149.

6 BSC, *The Return to Financial Viability: A Business Proposal for 1980–81*, Dec. 1979.

7 BSC, *Corporate Plan*, Dec. 1980.

8 The percentage share of employees in the iron and steel activities among total employees of the Corporation was 84.1 per cent in 1972 and 83.3 per cent in 1979 (BSC 1974, 1979/80).

9 The composition changes in employment over the period 1957–66 were as follows: administrative, technical and clerical employees increased from 14.9 per cent to 21.0 per cent, process workers decreased from 53.6 per cent to 46.2 per cent, general and maintenance employees increased from 31.5 per cent to 32.7 per cent (Hunter *et al.* 1970: 125).

10 See Note 4.

11 Industrial relations management increased its importance in function with

informal organisations at workplaces (organisations of workers led by shop stewards) in addition to trades unions and union leaders.

12 Nippon Steel Corporation offered technological assistance for the construction of the largest blast furnace of the Corporation; the Redcar blast furnace (4,750 m^3) began its operation in October 1979 (Nippon Steel 1981: 575).

13 The rate of dependence on imported iron ores was 49.3 per cent in 1960, 62.4 per cent in 1970 and 90.3 per cent in 1980. The dependence rate of coking coals was 5.6 per cent in 1980. The reason for increased foreign raw materials was that foreign ores and coals were superior in quality and lower in cost compared to the domestic ones (BSC 1978/79: 5, 16).

14 The category of manager includes managing directors, managers, management in special functions, scientists and technologists; that of supervisor includes functional specialists, foremen and clerical supervisors.

15 Figures for salaries of chairman and full-time board members and the top 0.7 per cent managers were calculated from annual emoluments. The number of managers included in the top 0.7 per cent was fifty-eight in 1972 and sixty-five in 1979 (BSC, op. cit., 1972/73, 1978/79).

9 MANAGEMENT AND LABOUR IN BRITISH STEEL WORKS

1 Another large-scale works of British Steel Corporation, Port Talbot Works, was rejected, although it is also a modernised works, mainly because it had not been equipped with continuous casters even in 1980. Continuous casters are one of the most important constituents of modernisation and rationalisation, just as LD converters are in the modern steel production system, for they rationalise ingot-making, soaking, and blooming or slabbing into one process, saving a great amount of production time, direct and indirect labour and costs. However, in order to achieve a more accurate comparison based upon similar technological conditions, Scunthorpe is later compared and reviewed with the existing large-scale works at Yawata in Japan in Chapter 11. In addition, a plant-level comparison is made for the blast furnace plant in Kure Works (BF No. 2) of the Nisshin Steel Co. Ltd, which has an almost identical blast furnace to Scunthorpe (BF No. 4), in order to supplement the comparison with Yawata, where a furnace of similar size was not in use when the research was undertaken.

2 Appleby–Frodingham Works (BSC), Scunthorpe Anchor Developments.

3 Occupations transferred to sub-contracting are: bricklayers (47 people), refractory main hands (30 people), steel service (24 people), scrap preparation (10 people), welfare attendants (4 people), office cleaners (one person) (Scunthorpe Works (BSC): Interview).

4 This marks a sharp contrast with modern Japanese works, where a young labour force of new graduates is specifically recruited (see Chapter 4, pp. 48–51 and Table 6.10).

5 Crafts unions which are members of the national Craft Co-ordinating Committee are as follows: ASBSBSW (Amalgamated Society of Boilermakers, Shipwrights, Blacksmiths and Structural Workers); EETPU (Electrical, Electronic, Telecommunication and Plumbing Union); AUEW–F

(Amalgamated Union of Engineering Workers–Foundry Section); AUEW–E (Amalgamated Union of Engineering Workers–Engineering Section); AUEW–TASS (Amalgamated Union of Engineering Workers–Technical Administrative and Supervisory Section); UCATT (Union of Construction, Allied Trades and Technicians); NUSMWCH & DE (National Union of Sheet Metal Workers, Coppersmiths, Heating and Domestic Engineers); APACS (Association of Patternmakers and Allied Craftspersons), BRTTS (British Roll Turners' Trade Society) (TUC: Interview).

6 In the reorganisation of September 1980 product-based businesses were established under three operating groups (BSC, *Annual Report and Accounts*: 1980/81).

7 The relative importance of industrial relations management in the British enterprises was identified by Brown (1981) as a general trend with more specialist industrial relations managers as early as 1978.

8 The case of comparison for the blast furnace plant is the Kure Works of Nisshin Steel Co. Ltd and for the LD converter plant the third steel plant of the Fukuyama Works of the Nippon Kōkan KK.

9 The seniority system includes, besides the one mentioned here: works seniority – often determined by the date of entry on the person's union membership card; departmental seniority – dependent on the length of time the worker has been employed in that particular department of the works; job seniority – dependent on the time employed at any particular job. It depends on the situation as to which seniority is preferred (ISTC: Interview).

10 NEBSS (National Examinations Board for Supervisory Studies) was established in 1964 on the initiative of the Department of Education and Science. It is composed of representatives of management and trades unions, and professional organisations. The BSC provides the education and training programme approved by this organisation. The majority of attendants are in the latter half of their thirties and the number of hours' training they receive is 240 (BSC: Interview).

11 The job content of foremen in Japanese works includes work management (30.9 per cent), labour management (16.4 per cent), safety management (13.1 per cent), education and training (12.6 per cent), self-management activity (10.4 per cent), cost control (10.3 per cent) and others (7.5 per cent) (Tani 1973).

12 The Japanese blast furnace work is usually divided into two teams: hob work and operation room work (hoisting, hot stove); and the LD converter work is divided into five teams: raw material, refining; casting, ingot-making, refining; continuous casting; crane; others (see Chapter 6).

10 MANAGEMENT AND LABOUR AT HEAD OFFICE (UK)

1 BSC, *Third Report on Organisation*, House of Commons Paper No. 60, HMSO, 1969b.

2 DTI, *British Steel Corporation: the Road to Viability*, March 1978, Cmnd 7149.

3 This objective is designed to be consistent with strengthening its marketing and technological position in the world steel industry; providing British

industry with products that are competitive in price, quality and service; and ensuring the efficient and socially responsible utilisation of human resources (BSC 1969).

4 BSC: Interview.

5 BSC: Interview.

6 The Corporation classifies manual workers into process operatives, maintenance and craftsmen and ancillary workers, and the rates of decrease were 31.1 per cent, 24.7 per cent and 22.5 per cent respectively, showing those of process operatives to be the largest (BSC: Interview).

7 BSC: Interview.

8 Here, management services, pensions office, shipping offices, records services, scrap buying offices, research laboratories, product advisory centres, product units and BSC (Industry) Ltd are included.

9 The Select Committee on Nationalised Industries 1977a: 11; Bank and Jones 1977: 9; Brannen *et al.* 1976: 96.

10 The number of unofficial strikes in the Corporation was 162 in 1971,171 in 1972, 251 in 1973, 345 in 1974 and 459 in 1975, thus showing an upward tendency (The Select Committee on Nationalised Industries 1977a: 20).

11 JAPAN/UK COMPARISON OF MANAGEMENT AND LABOUR

1 It was first recommended by the US government as early as 1946 but it was not until 1962 that the Yawata Steel Corporation became the first steel corporation to introduce it (Nippon Steel 1981: 652–3).

2 This nationalisation itself was a consequence of both idealism (socialist practice) and realism (necessity for modernisation) which affected the centralisation of the industrial organisation. In this sense it is contrary to the traditional British trend of decentralisation. In practice, however, they needed a decentralised corporate structure and decentralisation of management, which was also a requirement of passive compatibility in management at the declining phase of industrial society.

13 FURTHER CHANGES IN THE 1990S

1 The foreign exchange rate of the yen in April–June 1990 was ¥155/dollar while in the same period of 1993 it was ¥107/dollar (Tekkō Renmei 1993: 6).

2 I owe the information and knowledge in this section to an interview at Nippon Steel Corporation and Kobe Steel in April 1993. Although the plan has not yet been finalised, their new concept of business and general planning were hinting at change similar to that undergone by BSC.

3 In the case of Kobe Steel, they reformulated all corporate functions into three major groups: strategic function, common corporate function and horizontal function. In each diversified business, a self-contained administration will be set up, responsible for individual policy and personnel management (Tekkō Renmei, *Tekkō Kaihō 1994*: No. 1611).

4 Tekkō Renmei, *Tekkō Kaihō 1991*: Vol. 41, No. 11, 45–57.
5 The Japanese steel industry obtains raw materials almost entirely from overseas sources. The largest percentage of iron ore import (in 1992) is from Australia (46 per cent) followed by Brazil (23 per cent), India (15 per cent). The largest percentage of coal import (in 1992) is also from Australia (49 per cent) followed by Canada (23 per cent), USA (14 per cent) (Tekkō Renmei 1993).
6 Tekkō Renmei, *Tekkō Kaihō 1994*: No. 1616.
7 Nippon Steel: Interview.
8 The author owes information for this section to interviews made both at head office of British Steel plc and at ISTC.
9 The total number of major privatisations from 1979 to 1992 was 26 (*The Economist*, 1 January 1993) and in the year that British Steel was privatised, the Rover Group was also sold to British Aerospace, which was then, in January 1994, sold to BMW for £800 million (*The Economist*, 5 February 1994).
10 Aylen argues that even under public ownership efficiency can be and indeed was achieved by the BSC in the 1980s. Before 1980 enterprise efficiency was subordinated to immediate, short-term political concerns, while after 1980 commercial objectives were set, and social objectives were subordinated to the need for lower costs and financial viability (Aylen 1988: 3).
11 British Steel announced the agreement made with the owners of Walker to purchase their shares in October 1989. At this time British Steel, via the British Steel Distribution Division and Walker, were the two largest steelstock holders in the UK, accounting for 58 per cent of total purchases by the UK stockholders (Monopolies and Mergers Commission 1990).
12 The merger in July 1992 created Avesta Sheffield AB, British Steel holding 40 per cent of its equity. The new company is one of the largest European producers of cold-rolled stainless steel products with 600,000 tonnes of stainless steel and over 9,000 employees (British Steel, *Steel News* 6 July 1992).
13 In contrast to management assertions, the union opposed the closure not only because of the loss of over 700 jobs but also in view of the security of steel supply for manufacturing industries in Scotland (ISTC 1990, Rolling Mill Record, No.1).
14 In Kamaishi works, which was a small-scale old works, a new company which manufactures steel furniture was established in 1991 as a joint company of Nippon Steel and Okamura Seisakusho. It uses the renovated rolling mill plant and has 106 employees with a turnover of ¥1.1 billion (Nippon Steel *SHINNITETSU* 1992: Vol. 69).
15 In British Steel the union, ISTC, is much concerned with the issue of sub-contracting, regarding the working conditions of the same union members or potential members, and this union attitude is very different from that of the Japanese unions (ISTC, *Phoenix* August 1991).
16 This includes jobs such as safety, health, education/training and security, which implies a transfer of such functions to sub-contracting companies. But as sub-contracting companies cannot afford the same level of services, the cost will be borne by workers.
17 Ideological control of remaining employees would be enhanced as this kind

of division would foster among the remaining employees a sense of loyalty and superiority over sub-contract workers who are regarded as second-class citizens in the workplace.

18 The effect of technological innovation upon labour provided management with a chance to redesign the work in such a way as to gather some of the standardised tasks into a new general job group, so that workers can be more flexibly employed. This redesign has also helped to alleviate the monotony of doing simple standardised work. In this area of flexibility Japan was more advanced due to the non-job-specific type of employment.

19 I am indebted to the interview at ISTC (1993) for the information on industrial relations in the 1990s.

20 The difference in sick and injury allowances between white- and blue-collar employees is one of the important contemporary issues (ISTC: Interview).

Bibliography

JAPANESE

Chokki, A. (1983) *Gendai no Keiei Soshiki: Sono Kōzō to Dainamizumu* (Contemporary Managerial Organisation: its Structure and Dynamics), Tokyo: Yuhikaku.

Chujo, T. (1985) *Rōshi Kankei no Shiteki Kadai* (The Historical Theme of Industrial Relations), Tokyo: Chuokeizai-sha.

Daiyamondo-sha (1979) *Kaisha Shokuinronku* (Company Personnel Lists), Tokyo: Daiyamondo-sha.

Dore, R.P. (1980) 'Daikigyō to minshushugi' (Large-scale enterprise and democracy), *Nihon Rōdō Kyōkai Zassi*, Vol. 22, No. 10: 2–11.

—— (1984) 'Nijyu kōzō no saikentō' (Re-examination of double structure), *Nihon Rōdō Kyōkai Zassi*, Vol. 26, No. 4, 5: 9–17.

—— (1985) 'Chingin kōshō ni okeru kōsei no gainen' (The idea of fairness in wage bargaining), *Nihon Rōdō Kyōkai Zassi*, Vol. 27, No. 1: 4–13.

—— (1987) 'Preface to the Japanese edition' in *Igirisu no Kōjō – Nippon no Kōjō* (British Factory – Japanese Factory), Tokyo: Chikuma Shobo.

Fukada, Y. (1971) *Shinnitetsu no Teihen kara: Shitauke Rōdōsha no Hōkoku* (From the Base of Nippon Steel Corporation: Reports from Sub-contract Workers), Tokyo: San-ichi Shobo.

Hatano, T. (1987) *Tekkō Rōdō Tsushin* (No. 247), Tokyo: Tekkō Rōdō Tsushin.

Hazama, H. (1974) *Igirisu no Shakai to Rōshi Kankei* (British Society and Industrial Relations), Tokyo: Nihon Rōdō Kyōkai.

—— (1964) *Nihon Romukanrishi Kenkyū* (A Study of Japanese Labour Management History), Tokyo: Daiyamondo-sha.

Ichikawa, H. (1969) *Nihon Tekkōgyō no Saihensei* (The Reorganisation of the Japanese Iron and Steel Industry), Tokyo: Shinhyōron.

Iida, K. (1977) *Rōdō Undō no Tenkai to Rōshi Kankei* (The Evolution of Labour Movements and Industrial Relations), Tokyo: Mirai-sha.

Iida, S. (1970) *Nihon no Tekkō Shihon* (Capital of Japanese Iron and Steel), Tokyo: San-ichi Shobo.

Imada, O. (1981) 'Sagyo rōdōsha no rōdō to kanri (Labour of manual workers and management) in Ishida, K. (ed.) *Gendai Nihon no Tekkō Kigyō Rōdō* (Labour in the Modern Iron and Steel Industry), Kyoto: Minerva Shobo.

Inagami, T. (1984) 'Indasutoriarizumu to rōdō kenkyū' (A study of labour and industrialism), *Nihon Rōdō Kyōkai Zassi*, No. 300: 36–44.

Iron and Steel Statistics Bureau (1980) *Annual Statistics for the United Kingdom*, London: Iron and Steel Statistics Bureau.
Ishida, H. (1976) *Nihon no Rōshi Kankei to Chingin Kettei* (Japanese Industrial Relations and Determination of Wages), Tokyo: Toyokeizai Shinpo-sha.
Ishida, K. (1967) *Gendai Kigyō to Rōdō no Riron* (The Theory of Labour and Modern Enterprise), Kyoto: Minerva Shobo.
—— (1981) *Gendai Nihon no Tekkō Kigyō Rōdō* (Labour in the Contemporary Japanese Iron and Steel Enterprise), Kyoto: Minerva Shobo.
Ishida, M. (1982) 'Igirisu tekkogyo shinsetsu kōjō ni okeru rōshikosho no tenkai (Evolution of industrial relations in new plants of the British iron and steel industry), *Keizai Ronsō*, No. 130: 28–53.
Ito, K. (1994) 'Jinteki shigenkanri no tenkai' (Evolution of human resource management) in G. Sasagawa, T. Inamura and S. Inove (eds) *Gendai Keiei no Kihon Mondai* (Major Issues of Contemporary Management), Tokyo: Zeimukeirikyokai.
Itozono, T. (1978) *Nihon no Shagaiko Seidō* (The Japanese Sub-contracting System), Kyoto: Minerva Shobo.
Iwade, H. (1991) *Eikoku Rōmu Kanri* (Labour Management in Britain), Tokyo: Yuhikaku.
Iwai, M. (1981) *Shinnittetsu Manpawa Kakumei* (The Manpower Revolution of the Nippon Steel Corporation), Tokyo: Daiyamondo-sha.
Iwao, H. (1974) *Keiei Keizaigaku* (Business Economics), Tokyo: Maruzen.
Japan Institute of Labour (monthly) *Nippon Rōdō Kenkyu Zasshi* (Monthly Journal of the Japan Institute of Labour), Tokyo: Japan Institute of Labour.
Japan Institute of Labour (1993) '21 seiki ni muketeno nippon kigyō no jinzai ikusei' (Human resource development of Japanese enterprises towards the twenty-first century), *Nippon Rōdō Kenkyu Zasshi* (Journal of Japanese Labour Studies), Tokyo: No. 401: 2–60.
Jyukagakukogyo Tsushinsha (1974) *Gaikoku Gijyutsu Dōnyū Yōran* (Survey of Imported Technology) Tokyo: Jyukagakukōgyō Tsushin-sha.
Kagaku Gijyutsucho (1975) *Kagaku Gijyutsu Hakusho* (White Paper on Science and Technology) Tokyo: Kagakugijyutsuchō.
Kagono, T. (1980) *Keiei Soshiki no Kankyō Tekiyō* (Environmental Adaptation and Managerial Organisations), Tokyo: Hakuto Shobō.
Kato, M. (1981) 'Howaito kara no rōdō to kanri' (Labour and management of white collar employees) in K. Ishida (ed.) *Gendai Nippon no Tekkō Kigyo Rōdō* (Labour in Modern Iron and Steel Enterprises), Kyoto: Minerva Shobo.
Kawasaki Steel Corporation (1976) *Kawasaki Seitetsu 25 nen-shi* (Twenty-five years of Kawasaki Steel), Kobe: Kawasaki Steel.
—— (1977) *Yukashōken Hōkokusho* (Annual Report and Accounts), Tokyo: Okurasho.
Kawasaki, T. (1962) *Nihon Tekkōgyo no Hatten to Tokushitsu* (The Development and Characteristics of the Japanese Iron and Steel Industry), Tokyo: Kōgyo Tōsho Shuppan.
Keizai Kikauchō (Economic Planning Agency) (annual) *Keizai Hakusho* (Economic White Paper), Tokyo: Keizai Kikakuchō.
—— (annual) *Keizai Yōran* (Economic Summary) Tokyo: Keizai Kikakuchō.
Kimitsu Works (Nippon Steel) (1974) *Nyu Raifu in Kimitsu* (New Life in Kimitsu) Kimitsu: Nippon Steel.
—— (Nippon Steel) (1975) *Kimitsu Seitetsusho 10 nen no Ayumi* (Ten years of the Kimitsu Steel Works), Tokyo: Nippon Steel.

—— (Nippon Steel) (1985) *Kimitsu Seitetsusho Nijyunenshi* (Twenty years of the Kimitsu Steel Works), Kimitsu Works: Nippon Steel.

—— (Nippon Steel) (n.d.) *Kimitsu Seitetsusho Sōgō Jyōhōshori Shisutemu* (The Total Information System in Kimitsu Works) Kimitsu: Nippon Steel.

Kishino, H. (1977) *Nippon Kōkan: Gijitsu e no Chōsen* (Nippon Kōkan: Challenge to Technology), Tokyo: Union Shuppan-sha.

Kobe Steel Ltd (1974) *Kobe Seiko 70 nen* (Seventy years of Kobe Steel), Kobe: Kobe Steel, Ltd.

—— (1977) *Yukashōken Hōkokusho* (Annual Report and Accounts), Tokyo: Okurasho.

Koike, K. (1977) *Shokuba no Rōdō Kumiai to Sanka: Rōshi Kankei no Nichibei Hikaku* (Participation and Labour Unions at the Workshop Level: US–Japan Comparison in Industrial Relations), Tokyo: Tōyōkeizai Shinpo-sha.

—— (1978) *Rōdōsha no Keiei Sanka: Seiyo no Keiken to Nippon* (Workers' Participation: Experiences in Western Europe and Japan), Tokyo: Nihon Hyōron-sha.

Kurita, K. (1978) *Gendai Rōshi Kankei no Kōzō* (The Structure of Modern Industrial Relations), Tokyo: Tokyo Daigaku Shuppan.

Maeda, I. (1978) *Shinnitetsu, Chūgoku Kensetsu Tai* (Nippon Steel Corporation, Construction Team to China), Tokyo: Ko Shobō.

Mainichi Shimbun-sha (1969) *Shinnitetsu Tanjyōsu* (The Formation of the Nippon Steel Corporation), Tokyo: Mainichi Shimbun-sha.

Masaki, H. (1968) 'Tekkō dokusen ni okeru shikinchōtatsu no hensen to sono taiyō' (Collection of funds and means therefore by iron and steel monopolies) *Shakai Kagaku* (Social Science), Vol. 2, No. 2, Kyoto: Doshisha University.

Matsuzaki, T. (1982) *Nippon Tekkō Sangyō Bunseki* (An Analysis of the Japanese Iron and Steel Industry), Tokyo: Nihon Hyōron-sha.

Michimata, K. (ed.) (1978) *Gendai Nihon no Tekkō Rōdō Mondai* (Iron and Steel Labour Problems in Modern Japan), Hokkaido: Hokkaido Daigaku Tosho Kankō Kai.

Murakami, Y. (1975) *Sangyō Shakai no Byōri* (The Pathology of Industrial Society), Tokyo: Chuokōron-sha.

Naito, K. (1972) *Igirisu no Rodosha Kaikyu* (The Working Class in Britain), Tokyo: Tōyōkeizai Shinpo-sha.

Nakata, M. (1983) *Gendai Kigyō Kōzō to Kanri Kinō* (The Structure and Function of Modern Enterprise), Tokyo: Chuo Keizai-sha.

—— (1993) 'Kigyō kanri shisutemu no kokusaihikaku' (International comparison of management systems) in Scott, Nakata and Hasegawa *Kigyō to Kanri no Kokusai Hikaku* (International Comparison of Enterprise and Management), Tokyo: Chuo Keizai-sha.

Nakayama, I. (1960) *Nihon no Kōgyoka to Rōshi Kankei* (The Industrialisation of Japan and its Industrial Relations), Tokyo: Nihon Rōdō Kyōkai.

Nihon Ginkō (Bank of Japan) (annual) *Kokusai Hikaku Tokei* (International Statistics in Comparison), Tokyo: Nihon Ginkō Chōsa Tōkeikyoku, 1970/86.

Nippon Kōkan KK (1972) *Nippon Kōkan Kabushiki Kaisha 60 nen Shi* (Sixty years of Nippon Kōkan KK), Tokyo: Nippon Kōkan KK.

—— (1977) *Yukashōken Hōkokusho* (Annual Report and Accounts), Tokyo: Okurasho.

Nippon Steel Corporation (1970) *Hirohata Seitetsusho 30 nen Shi* (Thirty years of the Hirohata Steel Works), Tokyo: Nippon Steel.

—— (1971) *Chuo Kenkyūsho 10 nen Shi* (Ten years of the Central Research Institute), Tokyo: Nippon Steel.
—— (1975a) *Tekkō Rokusha Keiei Bunseki Kisō Shiryō* (Fundamental Documents for financial analysis of the Six Iron and Steel Corporations), Tokyo: Nippon Steel.
—— (1975b) *Hikari Seitetsusho 20 nen no Ayumi* (Twenty years of the Hikari Steel Works), Tokyo: Nippon Steel.
—— (1975c) *Chōsen no Hibi: Shinnitetsu no JK Katsudō kara* (Days of Challenge: From JK Activity of the Nippon Steel Corporation), Tokyo: Nippon Steel.
—— (1976) *Sengo 30 nen no Nippon Tekkōgyo no Hatten* (The Development of the Japanese Steel Industry over the last Thirty Years) Tokyo: Nippon Steel.
—— (1978a) *Obei Tekkōgyo no Shomondai* (Problems of the US and European Steel Industries), Tokyo: Nippon Steel.
—— (1978b) *Wagasha no Gaiyō* (Outline of our Company), Tokyo: Nippon Steel.
—— (1980) *Yawata Seitetsusho 80 nen Shi:* 4 vols. (Eighty years of the Yawata Steel Works), Tokyo: Nippon Steel.
—— (1981a) *Honō to Tomoni*, 3 vols. (Friends of Fire, 3 vols.), Tokyo: Nippon Steel.
—— (1981b) *Shashi Bessatsu Sankō Shiryōshū* (1950–80) (Reference Documents on the History of the Nippon Steel Corporation), Tokyo: Nippon Steel.
—— (annual) *Yukashōken Hōkokusho* (Annual Report and Accounts), Tokyo: Okurasho.
—— (annual) *Tōkei Yōran* (Statistical Handbook), Tokyo: Nippon Steel.
—— (monthly) *SHINNITETSU*, Tokyo: Nippon Steel.
Nippon Steel Kimitsu Labour Union (1980) *Shinnitetsu Kimitsu Rōdō Undōshi* (The History of Nippon Steel Kimitsu Labour Union), Kimitsu: Nippon Steel Kimitsu Labour Union.
Nippon Steel Labour Union Federation (1982) *Shinnitetsu Rōdō Undōshi* (The History of the Labour Movement in the Nippon Steel Corporation), Tokyo: Nippon Steel Labour Union Federation.
Nosaka, Y. (1970) *Tekkōgyo no Computer Control* (Computer Control in the Iron and Steel Industry), Tokyo: Sangyō Tōsho.
Okishio, N. and Ishida, K. (1981) *Nihon no Tekkōgyo* (The Japanese Iron and Steel Industry), Tokyo: Yuhikaku.
Okochi, K. (1955) *Sengo Nihon no Rōdō Undō* (The Labour Movement of Japan in the Post-war Period), Tokyo: Iwanami Shoten.
—— (ed.) (1956) *Rōdō Kumiai no Seisei to Soshiki* (The Formation and Structure of Trade Unions), Tokyo: Tokyo Daigaku Shuppan Kai.
Orii, H. (1973) *Rōmukanri Nijūnen: Nippon Kōkan ni miru Sengo Nihon no Rōmukanri* (Twenty Years of Labour Management: the Post-war Japanese Labour Management of Nippon Kokan, KK), Tokyo: Tōyōkeizai Shinpo-sha.
Peace Economic Planning Council (1978) *Kokumin no Dokusen Hakusho: Kigyō Shūdan* (The National Monopoly White Paper: Enterprise Group), Tokyo: Ochanomizu Shōbō.
Rōdō Tsushin (1987) *Tekkō Rōdō Tsushin* (Iron and Steel Correspondence), No. 247, Tokyo: Tekkō Rōdō Tsushin.

Rōdōsho (Ministry of Labour) (1976) *Shōgaikoku no Rōshi Kyōgisei to Keiei Sanka* (Participation and Industrial Relations in Foreign Countries), Tokyo: Nihon Rōdō Kyōkai.

Rōdōsho (annual) *Rōdō Hakusho* (White Paper on Labour), Tokyo: Nihon Rōdō Kyōkai, 1960–79.

Sasagawa, G. and Ishida, K. (eds) (1984) *Gendai Kigyō no Howaito Kara Rōdō* (White-Collar Labour in Modern Enterprises), Tokyo: Otsuki Shoten.

Shimada, E. (1970) *Oshu Tekkōgyo no Shuchu to Dokusen* (Concentration and Monopoly in the European Iron and Steel Industry), Tokyo: Shin Hyōron.

Shioda, N. (1973) *Sekai no Tekkō Shijo* (World Steel Markets), Tokyo: Shin Hyōron.

Sohyō (General Council of Trade Unions of Japan) (1980) *Hachijunendai Rōdō Undō no Tenbo* (Prospects for the Labour Movement in the 1980s), Tokyo: Rōdō Kyōiku Centre.

Sohyō (monthly) *Chōsa Geppō* (Monthly Survey), Tokyo: Sohyō.

Sorifu Tōkeikyoku (Statistical Bureau of Prime Minister's Office) (annual) *Kagakugijitsu Kenkyū Chosa Hōkoku* (Report on the Survey of Research and Development in Japan), Tokyo: Sorifu Tōkeikyoku.

Sumitomo Metal Industries Ltd (1967) *Sumitomo Kinzoku Kōgyō Saikin Jyunenshi* (The Last Ten Years of Sumitomo Metal Industries Ltd), Osaka: Sumitomo Metal Industries Ltd.

—— (1977a) *Sumitomo Kinzoku Kōgyō Saikin Jyunenshi* (The Last Ten Years of Sumitomo Metal Industries, Ltd), Osaka: Sumitomo Metal Industries Ltd.

—— (1977b) *Yukashōken Hōkokusho* (Annual Report and Accounts), Tokyo: Okurasho.

Sumiya, M. (1964) *Nihon no Rōdō Mondai* (Japanese Labour Problems), Tokyo: Tokyo Daigaku Shuppan Kai.

—— (ed.) (1967) *Tekkōgyō no Keizai Riron* (The Economic Theory of the Iron and Steel Industry), Tokyo: Nihon Hyōron-sha.

—— (1977) *Nihon Rōshi Kankei Shiron* (On the History of Industrial Relations in Japan), Tokyo: Tokyo Daigaku Shuppan Kai.

Tabe, S. (1969) *Tekkō Genryō Ron* Vol. 2 (On the Raw Materials for Iron and Steel) Tokyo: Diayamondo-sha.

Takahashi, K. (1978) *Nihonteki Rōshikankei no Kenkyu* (A Study of Japanese Industrial Relations), Tokyo: Mirai-sha.

Takahashi, T. (1967) *Igirisu Tekkō Dokusen no Kenkyu* (A Study of British Steel Monopolies), Kyoto: Minerva Shobo.

Takanashi, A. (1967) *Nihon Tekkōgyo no Rōshi Kankei* (Industrial Relations in the Iron and Steel Industry of Japan), Tokyo: Tokyo Daigaku Shuppan Kai.

Tanabe, K. (1981) *Tekkōgyo* (The Iron and Steel Industry), Tokyo: Tōyōkeizai Shinpo-sha.

Tani, E. (1973) 'Tekkōgyo no sagyōchō' (Foremen in the steel industry) in Tekkō Renmei (monthly) *Tekkō no IE*, May 1973.

Tekko Kyokai (The Iron and Steel Institute of Japan) (1972) *Tekkō Seizōhō* (Iron and Steel Production Methods) Vol.1, Tokyo: Maruzen.

—— (1974) *Nishiyama Gijyutsu Kōza* (Nishiyama Memorial Lecture) Tokyo: Tekkō Kyōkai.

—— (1975) *Tetsu to Hagane* (Iron and Steel) Tokyo: Tekko Kyōkai.

—— (1976) *Wagakunini okeru Saikin no Hot-strip Setsubi oyobi Seizo Gijyutsu no Shinpo* (Recent Progress in Hot-strip Mills and Production Technology) Tokyo: Tekkō Kyōkai.

—— (1977) *Wagakunini okeru Seisen Gijyutsu no Shinpo* (Progress in Ironmaking Technology in Japan) Tokyo: Tekkō Kyōkai.
—— (1982) *Tekkōgyo ni okeru Compyuta Riyo no Genjō to Tenbō* (The Prospect and Present Situation of the Use of Computers in the Iron and Steel Industry), Tokyo: Iron and Steel Institute of Japan.
Tekkō Renmei (The Japan Iron and Steel Federation) (1959) *Sengo Tekkōshi* (The Post-war History of Iron and Steel), Tokyo: Tekkō Renmei.
—— (1962) *Tekkō Kakusha Kyōiku Taikei, Kiteishu* (Iron and Steel Corporations' Education Systems and Rules), Tokyo: Tekkō Renmei.
—— (1966) *Tekkō Kakusha ni okeru Keieisha, Kanrisha Kyōiku* (Education of Administrators and Managers of Iron and Steel Corporations), Tokyo: Tekkō Renmei.
—— (1968) *Tekkōgyo ni okeru IE Katsudō Jyunenshi* (Ten Years of IE Activity in the Iron and Steel Industry), Tokyo: Tekkō Renmei.
—— (1969) *Tekkō Jyunenshi* (1958–67) (Ten Years of Iron and Steel), Tokyo: Tekkō Renmei.
—— (1972) *Tekkōgyo Kanrisha no Nōryoku Kaihatsu* (Faculty Development of Managerial Personnel in the Iron and Steel Industry), Tokyo: Tekkō Renmei.
—— (1974a) *Obei Tekkōgyo no Rōshi Kankei to Chingin Jijyo* (Industrial Relations and the Situation of Wages in the Iron and Steel Industry of the US and Europe), Tokyo: Tekkō Renmei.
—— (1974b) *Obei no Koyō Jijyo* (The Employment Situation in the US and Europe), Tokyo: Tekkō Renmei.
—— (1977) *Tekkō Kakusha Nōryoku Kaihatsu Keikaku* (The Programme of Faculty Development in the Iron and Steel Corporations), Tokyo: Tekkō Renmei.
—— (1979) *Sōzō to Hatten no Jyunen: Tekkogyō ni okeru Jishukanri Katsudō no Ayumi* (Ten Years of Creation and Development: The Evolution of Self-management Activity in the Iron and Steel Industry), Tokyo: Tekkō Renmei.
—— (1981) *Tekkō Jyunenshi* (1968–77) (Ten Years of Iron and Steel: 1968-77), Tokyo: Tekkō Renmei.
—— (1985) *Nihon Tekkōgyo ni okeru Jishu Kanri Katsudō* (Jishu Kanri Katsudō in the Japanese Steel Industry) Tokyo: Tekkō Renmei.
—— (1994) *1993 nen no Naigai Tekkōgyo* (The Steel Industry in Japan and Abroad in 1993), Tokyo: Tekkō Renmei.
—— (annual) *Tekkō Tōkei Yōran* (Iron and Steel Statistical Handbook), Tokyo: Tekkō Renmei.
—— (annual) *Nihon no Tekkōgyo* (The Japanese Steel Industry) Tokyo: Tekkō Renmei.
—— (monthly) *Tekkō Kaihō* (The Iron and Steel World), Tokyo: Tekkō Renmei.
—— (monthly) *Tekkō no IE* (Industrial Engineering for Iron and Steel) Tokyo: Tekkō Renmei.
Tekkō Rōren (Japanese Federation of Iron and Steel Workers' Unions) (1971) *Tekkō Rōdō Undōshi* (The History of the Iron and Steel Labour Movement), Tokyo: Tekkō Rōren.
—— (1975) *Oote Seitetsusho no Shitauke Kigyō niokeru Rōshi Kankei* (Industrial Relations in Sub-contract Firms in Large-scale Iron and Steel Works), Tokyo: Tekkō Rōren.

—— (1977) *Ōte Seitetsusho no Rōdōsha to Rōdō Kumiai* (Workers and Labour Unions in Large-scale Iron and Steel Works), Tokyo: Tekkō Rōren.
—— (annual) *Kanren Rōdōsha Handbook* (Affiliated Workers' Handbook), Tokyo: Tekkō Rōren.
—— (annual) *Tekkō Rōdō Handbook* (Iron and Steel Handbook), Tokyo: Tekkō Rōren.
Tekkō Rōren and Rōdō Chōsa Kyōgikai (Labour Research Council) (ed.) (1980) *Tekkō Sangyō no Rōshi Kankei to Rōdō Kumiai* (Labour Unions and Industrial Relations in the Iron and Steel Industry), Tokyo: Nihon Rōdō Kyōkai.
Tekkō Shimbun-sha (annually) *Tekkō Nenkan* (Iron and Steel Yearbook), Tokyo: Tekkō Shimbun-sha.
Thurley, K.E. (1970) 'Igirisu rōshi kankei no genkyo: Donovan hōkoku to sono gō' (The current situation of industrial relations in Britain: The Donovan report and thereafter), *Nihon Rōdō Kyōkai Zassi*, No. 140: 10–18.
—— (1984) 'Igirisu kara mita Nihon no rōshi kankei' (Japanese industrial relations: a British perspective), *Nihon Rōdō Kyōkai Zassi*, No. 300: 78–84.
Toda, T (1984) *Gendai Sekai Tekkōgyo Ron* (A Study of Contemporary World Steel Industry), Tokyo: Bushindo.
Tokyo Shoko Research (1978) *Tosho Shinyo Roku* (Tosho Trust Handbook) Tokyo: Tokyo Shoko Research
Tokyo University Social Science Research Institute (1985) *Igirisu Rōshi Kankei no Jittai Chōsa*, No. 2. (Survey of Industrial Relations in Britain, No. 2), Research Report No. 20, Tokyo: The Tokyo University Social Science Research Institute.
Tominaga, K. (1967) 'Rōshikankei bunseki no shakaigakuteki kiso' (The sociological foundation for the analysis of industrial relations), *Nihon Rōdō Kyōkai Zassi*, No. 100.
Tomizawa, K. (1978) *Rōdō to Dokusen Shihon* (Labour and Monopoly Capital), Trans. Harry Braverman (1974) Tokyo: Iwanami Shoten.
Toyama, Y. (1973) *Igirisu Sangyō Kokuyuka Ron* (On Nationalisation of the British Industry), Kyoto: Minerva Shobo.
Tōyō Keizai (1975) *Kigyō Keiretsu Soran* (Prospectus Regarding Affiliation of Enterprises), Tokyo: Tōyō Keizaishinpo-sha.
Trevor, M. (1988) *Eikoku Toshiba no Keiei Kaushin* (Toshiba's New British Company), Tokyo: Tōyōkeizai Shinpo-sha.
Tsuda, M. (1973) *Shudan Shugi Keiei no Kōsō* (The Conception of Collective Management), Tokyo: Sangyo Rōdō Chōsa-shō.
—— (1977) *Nihon no Rōmu Kanri* (Japanese Labour Management), Tokyo: Tokyo Daigaku Shuppan Kai.
Tsūsanshō (MITI, The Ministry of International Trade and Industry) (annual) *Shuyo Sangyō no Setsubitoshi Keikaku* (Plan of Investment for Plant Equipment by Major Industries) Tokyo: Tsūsanshō.
—— (1973a) *70 nendai no Tekkōgyo* (The Iron and Steel Industry in the 1970s), Tokyo: Tsūsanshō.
—— (1973b) *Minkan Setsubi Toshi no Chuki Tenbo* (Middle Term Prospects for Private Investment in Equipment) Tokyo: Tsūsanshō.
Uchida, T. (1969) *Yoroppa Keizai to Igirisu* (The European Economy and Britain), Tokyo: Tōyōkeizai Shinpo-sha.
Ujihara, S. (1966) *Nihon Rōdō Mondai Kenkyū* (A Study of Japanese Labour Problems), Tokyo: Tokyo Daigaku Shuppan Kai.

Urabe, T. (ed.) (1979) *Soshiki no Kontinjanshi Riron* (The Contingency Theory of Organisation), Tokyo: Hakuto Shobo.
Wakayama Works (1977) *Wakayama Seitetsusho Brochure* (Wakayama Works), Wakayama: Sumitomo Metal Industries Ltd.
Watanabe, M. (1978) Book review in *Keizai Kenkyū*, Tokyo: 1978.
Watanabe, T. (1984) *Kigyō Kanri to Kanri Rōdō* (The Managerial Role in Business Management), Tokyo: Chikura Shobo.
Yamaguchi, M. (1979) Book review in *Keizai*, Tokyo: Shinnippon Shuppan.
Yamamoto, S. (1974) *Igirisu Sangyō Kōzō Ron* (On the Industrial Structure of Britain), Kyoto: Minerva Shobo.
Yamashita, H. (1977) 'Yoin gorika no hoho' (An approach to personnel rationalisation) *Tekkō no IE*, Tokyo: Tekkō Renmei.
Yasui, T. (1986) *Gendai Daikogyō no Rōdō to Kanri* (Labour and Management in Modern Large-scale Industries), Kyoto: Minerva Shobo.
Yawata Steel Corporation (1961) *Yawata Seitetsu Sanko Tōkei* (Statistical Reference Data for Yawata Steel) Yawata: Yawata Steel Corporation.
Yawata Works (1977) *Yawata Seitetsusho Brochure* (Yawata Works), Yawata: Nippon Steel.
Yoneyama, K. (1978) *Gijitsu Kakushin to Shokuba Kanri* (Technical Innovation and Workshop Management), Tokyo: Mokutaku-sha.
Yoshitake, K. (1968) *Igirisu Sangyō Kokuyuka Seisaku Ron* (A Study of the Nationalisation Policy of the British Industry), Tokyo: Nihon Hyōron-sha.

ENGLISH

Abegglen, J.C. (1973) *Management and Worker: The Japanese Solution*, Tokyo: Sophia University, Kodansha International.
Abromeit, H. (1986) *British Steel*, Heidelberg: Berg Publishers Ltd.
Allen, G.C. (1965) *Japan's Economic Expansion*, London: Oxford University Press.
—— (1978) *How Japan Competes: A Verdict on 'Dumping'*, London: The Institute of Economic Affairs.
—— (1979) *The British Disease*, London: The Institute of Economic Affairs.
—— (1981) *The Japanese Economy*, London: Weidenfeld and Nicolson.
Ansoff, H.I. (1965) *Corporate Strategy*, New York: McGraw-Hill.
Appleby–Frodingham Works (BSC) *Scunthorpe Anchor Developments*, Scunthorpe: BSC.
Armstrong, P. (1988) 'Labour and monopoly capital', in R. Hyman and W. Streeck (eds) *New Technology and Industrial Relations*, Oxford: Basil Blackwell.
Aylen, J. (1977) 'The British steel corporation and technical change' in *SCNI*, London: HMSO.
—— (1980) 'Innovation in the British steel industry' in K. Pavitt (ed.) *Technical Innovation and British Economic Performance*, London: Macmillan.
—— (1982) 'Plant size and efficiency in the steel industry: an international comparison', *National Institute Economic Review*, No. 100: 65–76.
—— (1984) 'Prospect for steel', *Lloyds Bank Review*, No.152: 13–30.
—— (1988) 'Privatisation of the British Steel Corporation' in *Fiscal Studies*, August: 1–25.

Babbage, C. (1835) *On the Economy of Machinery and Manufactures*, London: Charles Knight.
Bain, G.S. (1970) *The Growth of White-Collar Unionism*, Oxford: Clarendon Press.
Bain, J.S. (1959) *Industrial Organisation*, New York: John Wiley & Sons, Inc.
Bank, J. and Jones, K. (1977) *Worker Directors Speak*, Hants: Gower Press.
Barnard, C.I. (1938) *The Functions of the Executives*, Cambridge, Mass.: Harvard University Press.
Barnett, D.F. and Schorsch, L. (1983) *Steel*, Cambridge, Mass.: Ballinger.
Baumol, W.J. (1959) *Business Behavior, Value and Growth*, New York: Macmillan.
Bell, D. (1960) *The End of Ideology*, Glencoe, IU.: Free Press.
Bennis, W. and Mcgregor C. (eds) (1967) *The Professional Manager*, New York: McGraw-Hill.
Berle, A.A. and Means, G.C. (1932) *The Modern Corporation and Private Property*, New York: Macmillan.
Blauner, R. (1964) *Alienation and Freedom*, Chicago: University of Chicago Press.
Brannen, P., Batstone, E., Fatchett, D. and White, P. (1976) *The Worker Directors*, London: Hutchinson.
Braverman, H. (1974) *Labor and Monopoly Capital*, New York: Monthly Review Press.
Bright, J.R. (1955) 'How to evaluate automation', *Harvard Business Review*, Vol. 33, No. 4.
—— (1958) 'Economic aspects of automation', in *Proceedings of the EIA Conference on Automation Systems for Business and Industry*, New York: Engineering Publishers.
British Steel *Steel News*, 6 July 1992: British Steel plc.
British Steel Corporation (BSC) (1967) *Report on Organisation*, London: BSC.
—— (1969a) *Second Report on Organisation*, London: HMSO.
—— (1969b) *Third Report on Organisation*, London: HMSO.
—— (1975) *Organisation Review 1975*, London: HMSO.
—— (1979) *A Business Proposal for 1980/81*, London: BSC in ISTC (1980) *New Deal for Steel*, London: ISTC.
—— (1980) *Corporate Plan*, London: BSC.
—— (Annual) *Annual Report and Accounts*, London: BSC.
—— (Annual) *Annual Statistics for the Corporation*, London: BSC.
—— (International) Ltd (1979/80) *Report and Accounts*, London: BSC (International) Ltd.
Brown, A.W. (ed.) (1981) *The Changing Contours of British Industrial Relations*, Oxford: Basil Blackwell.
Bryer, R.A., Brignall, T.J. and Maunders A.R. (1982) *Accounting for British Steel*, Hants: Gower.
BSC/TUCSICC (1975) *JAPAN, Report of a Joint Visit*, London: BSC/TUCSICC.
Burn, D. (1961) *The Steel Industry 1939–59*, Cambridge: Cambridge University Press.
Burnham, J. (1941) *The Managerial Revolution*, New York: John Day.
Burns, T. and Stalker, G.M. (1961) *The Management of Innovation*, London: Tavistock Publications.
Caves, R. (1964) *Structure, Conduct, Performance*, New York: Prentice-Hall.
Central Office of Information (1975) *Steel*, London: HMSO.

Central Statistical Office (CSO) (annual), *Annual Abstract of Statistics*, London: HMSO.

Chandler, Jr. A.D. (1977) *The Visible Hand: The Managerial Revolution in American Business*, Cambridge, Mass.: Harvard University Press.

Clark, R. (1979) *The Japanese Company*, New Haven: Yale University Press.

Clarke, R.O., Fatchett, D.J. and Roberts, B.C. (1972) *Workers' Participation in Management in Britain*, London: Heinemann.

Clegg, H. (1970) *The System of Industrial Relations in Great Britain*, Oxford: Basil Blackwell.

—— (1979) *The Changing System of Industrial Relations in Great Britain*, Oxford: Basil Blackwell.

Cmnd 2651 (1965) *Steel Nationalisation*, London: HMSO.

Cmnd 3362 (1967) *British Steel Corporation: Report on Organisation*, London: HMSO.

Cmnd 3623 (1968) *Report of the Royal Commission on Trade Unions and Employers Associations*, London: HMSO.

Cmnd 5399 (1973) *British Steel Corporation*, London: HMSO.

Cmnd 7188 (1978) *British Steel Corporation*, London: HMSO.

Coates, D. and Hillard, J. (1986) *The Economic Decline of Modern Britain*, Sussex: Wheatsheaf Books.

Coates, K. (ed.) (1968) *Can the Workers Run Industry*?, London: Sphere Books.

Cockerill, A. (1974) *The Steel Industry: International Comparisons of Industrial Structure and Performance*, Cambridge: Cambridge University Press.

—— (1980) 'Steel', in P.S. Johnson (ed.), *The Structure of British Industry*, London: Granada.

Cole, R.E. (1979) *Work, Mobility and Participation: A Comparative Study of American and Japanese Industry*, Berkeley: University of California Press.

Commission of the European Communities (1982) *The European Steel Policy*: CEC.

Dahrendorf, R. (1959) *Class and Class Conflict in Industrial Society*, London: Routledge & Kegan Paul.

Department of Employment (1971) *British Labour Statistics: Historical Abstract 1886–1968*, London: HMSO.

Department of Trade and Industry (DTI), Cmnd 5226 (1973) *British Steel Corporation: Ten Year Development Plan*, London: HMSO.

—— Cmnd 7149 (1978) *British Steel Corporation: The Road to Viability*, London: HMSO.

Doeringer, P.B. and Piore, M.J. (1971) *Internal Labor Markets and Manpower Analysis*, Lexington, Mass.: Heath.

Dore, R.P. (1973) *British Factory – Japanese Factory*, London: Allen & Unwin.

—— (1983) *Flexible Rigidities: Industrial Policy and Structural Adjustment in the Japanese Economy*, 1970–80, Geneva: ILO.

Dunlop, J.T. (1958) *Industrial Relations Systems*, New York: Henry Holt & Co.

Dunning, J.H. (1986) *Japanese Participation in British Industry*, London: Croom Helm.

Economist, The 1 January 1993.

Economist, The 5 February 1994.

Economist, The 15 May 1994.

Elger, T. and Smith, C. (eds) (1994) *Global Japanization*?, London: Routledge.

Etzioni, A. (1964) *Modern Organizations*, Englewood Cliffs, N.J.: Prentice-Hall.
Evans, R. Jr. (1971) *The Labor Economies of Japan and the United States*, New York: Praeger Publishers.
Federal Trade Commission (Bureau of Economics) (1977) *The United States Steel Industry and its International Rivals*, Federation Trade Commission.
Flanders, A. (1965) *Industrial Relations: What is wrong with the system?*, London: Faber.
Foreman-Peck, J. and Hasegawa, H. (1989) 'Convergence theory and the development of the British and Japanese steel industries, 1960–1988', *Japan Forum*, Vol. 1, No.1: 29–41.
Forslin, J., Sarapata, A., Whitehill, A.M. (eds) (1979) *Automation and Industrial Workers*, Oxford: Pergamon Press.
Friedman, A.L. (1977) *Industry and Labour*, London: Macmillan.
Galbraith, J.K. (1978) *The New Industrial State* (Third Edition), Boston: Houghton Mifflin.
Garrahan, P. and Stewart, P. (1992) *The Nissan Enigma*, London: Mansell.
Gerschenkron, A. (1962) *Economic Backwardness in Historical Perspective*, Cambridge, Mass.: Harvard University Press.
Goldthorpe, J. H., Lockwood, D., Bechhofer, F. and Platt, J. (1968) *The Affluent Worker: Industrial Attitudes and Behaviour*, Cambridge: Cambridge University Press.
Hanami, T. (1979) *Labor Relations in Japan Today*, Tokyo: Kodansha International.
Hanson, C., Jackson, S., Miller, D. (1982) *The Closed Shop*, Hants: Gower.
Hartley, J., Kelly, J. and Nicholson, N. (1983) *Steel Strike: A Case Study in Industrial Relations*, London: Batsford Academic and Educational Ltd.
Hasegawa, H. (1988) *A Comparative Study of the Steel Industry in the UK and Japan*, PhD thesis submitted to the University of Newcastle upon Tyne.
Hasegawa, H. (1993) 'Japanese employment practices and industrial relations', *Japan Forum* Vol. 5, No.1: 21–35.
Hayek, F.A. (1944) *The Road to Serfdom*, London: George Routledge & Sons.
Hirschmeier, J. and Yui, T. (1981) *The Development of Japanese Business 1600–1980*, London: Allen & Unwin.
Hogan, W.T. (1983) *World Steel in the 1980s*, Toronto: Lexington Books.
Hudson, R. and Sadler, D. (1989) *The International Steel Industry*, London: Routledge.
Hunter, L.C., Reid, G.L. and Boddy, D. (1970) *Labour Problems of Technological Change*, London: George Allen & Unwin.
Iron and Steel Board (1964) *Development in the Iron and Steel Industry* (Special Report, 1964), London: HMSO.
—— (annual) *Annual Report*, London: The Iron and Steel Board.
Iron and Steel Institute (1963) *A Report by A Delegation from the Iron and Steel Institute: the Iron and Steel Industry of Japan*, London: The Iron and Steel Institute.
IISI (International Iron and Steel Institute) (1972) *Projection 85*, Brussels: IISI.
ILO (International Labour Organisation) (1992) *Recent Developments in the Iron and Steel Industry*, Geneva: ILO.
Industry and Trade Committee (ITC) (1981) *Effects of BSC's Corporate Plan*, London: HMSO (HC. 336–1, 336-ii of 1980/81, 2 vols).
—— (1982) *British Steel Corporation*, London: HMSO (HC 308 of 1981/82).

ISTC (The Iron and Steel Trades Confederation) (1980) *New Deal for Steel*, London: ISTC.

ISTC (1990) *Rolling Mill Record*, No.1.

ISTC (1991) *Phoenix*, No. 19.

Kawahito, K. (1972) *The Japanese Steel Industry*, New York: Praeger Publishers.

Kerr, C. (1983) *The Future of Industrial Societies: Convergence or Continuing Diversity*?, Cambridge, Mass.: Harvard University Press.

Kerr, C., Dunlop, J.T., Harbison F.H. and Myers, C.A. (1960) *Industrialism and Industrial Man: The Problems of Labor and Management in Economic Growth*, Cambridge, Mass.: Harvard University Press.

Kono, T. (1984) *Strategy & Structure of Japanese Enterprises*, London: Macmillan.

Lawrence, P.R. and Lorsh, J.W. (1967) *Organization and Environment: Managing Differentiation and Integration*, Boston, Mass.: Harvard Business School, Division of Research.

Leibenstein, H. (1976) *Beyond Economic Man*, Cambridge, Mass.: Harvard University Press.

Liberal Industrial Inquiry (1928) *Britain's Industrial Future*, London: Ernest Benn Ltd.

Littler, C.R. (1982) *The Development of the Labour Process in Capitalist Societies*, London: Heinemann Educational Books.

Lockwood, D. (1958) *The Black-Coated Worker*, London: Allen & Unwin.

March, J.G. and Simon, H.A. (1958) *Organizations*, New York: Wiley.

March, R.M. and Mannari, H. (1976) *Modernization and the Japanese Factory*, Princeton, NJ: Princeton University Press.

Marris, R. (1964) *The Economic Theory of 'Managerial' Capitalism*, London: Macmillan.

Marsh, A. (1979) *Concise Encyclopaedia of Industrial Relations*, Hants: Gower Press.

Marsh, A. and Ryan, V. (1984) *Historical Directory of Trade Unions* (Vol. 2), Hants: Gower.

Marshall, A. (1916) *Principles of Economics* (1890), London: Macmillan.

Marx, K. (1965) *Capital* (Vol. 2, 1867), Moscow: Progress Publishers.

McGregor, D. (1960) *The Human Side of Enterprise* (25th anniversary printing,1985), New York: McGraw-Hill Book Company.

Metal Bulletin Boots (1978) *Iron and Steel Works of the World*, 7th edn, Surrey: Metal Bulletin Books.

Monopolies and Mergers Commission, Cm 437 (1990) *British Steel Corporation*, London: HMSO.

Morishima, M. (1982) *Why has Japan 'Succeeded'?*, Cambridge: Cambridge University Press.

Nakayama, I. (1975) *Industrialization and Labor–Management Relations in Japan*, Tokyo: Japan Institute of Labour.

National Economic Development Office (NEDO) (1977) *A Study of UK Nationalised industries*, London: NEDO.

—— (1979) *Iron and Steel SWP* (Progress Report), London: NEDO.

—— (1980) *Iron and Steel SWP* (Progress Report), London: NEDO.

—— *A Hard Look at Steel (1, 2, 3)*, London: NEDO.

Nippon Steel (annual) *Basic Facts about Nippon Steel*, Tokyo: Nippon Steel.

Nishiyama, C. and Allen, G.C. (1974) *The Price of Prosperity*, London: The Institute of Economic Affairs.

Organisation for Economic Cooperation and Development (OECD) (1973) *Manpower Policy in JAPAN*, Paris: OECD.

—— (1977) *The Development of Industrial Relations Systems: Some Implications of Japanese Experience*, Paris: OECD.

Okochi, K., Karsh, B., Levine, S.B. (eds) (1974) *Workers and Employers in Japan: The Japanese Employment Relations System*, Princeton, NJ: Princeton University Press.

Oliver, N. and Wilkinson, B. (1988; 2nd edn 1992) *The Japanization of British Industry*, Oxford: Basil Blackwell.

Olson, M. (1982) *The Rise and Decline of Nations: Economic Growth, Stagflation and Social Rigidities*, New Haven and London: Yale University Press.

Ovenden, K. (1978) *The Politics of Steel*, London: Macmillan.

Ozaki, M. *et al.* (1992) *Technological Change*, Geneva: ILO.

Reischauer, E.O. (1977) *The Japanese*, Cambridge, Mass.: Harvard University Press.

Rostow, W.W. (1960) *The Stages of Economic Growth*, Cambridge: Cambridge University Press.

Rowley, C.K. (1971) *Steel and Public Policy*, New York: McGraw-Hill.

—— (1983) 'The political economy of the public sector', in J.B. Barry Jones (ed.) *Perspectives on Political Economy*, London: Francis Pinter.

—— (ed.) (1972) *Readings in Industrial Economics*, Vol. 1 and 2, London: Macmillan.

Rowley, C.K. and Peacock, A.T. (1975) *Welfare Economics*, London: Martin Robertson.

Scherer, F.M. (1970) *Industrial Market Structure and Economic Performance*, Chicago: Rand McNally.

Schumpeter, J.A. (1950) *Capitalism, Socialism, and Democracy*, New York: Harper.

Select Committee on Nationalised Industries (SCNI) (1973) *The British Steel Corporation*, London: HMSO.

—— (1977a) *The British Steel Corporation First Report*, London: HMSO.

—— (1977b) *The British Steel Corporation*, London: HMSO.

—— (1978) *Financial Forecasts of the British Steel Corporation*, London: HMSO.

Scott, J. (1986) *Capitalist Property & Financial Power*, Brighton: Wheatsheaf Books.

Scott, W.H., Halsey, A.H., Banks, J.A. and Lupton, T. (1956) *Technical Change and Industrial Relations: A study of the relations between technical change and the social structure of a large steel works*, Liverpool: Liverpool University Press.

Silberston, A. (1978) 'Nationalised industries: government intervention and industrial efficiency' in D. Butler, and A.H. Halsey (eds) *Policy & Politics*, London: Macmillan.

—— (1982) 'Steel in a mixed economy', in Lord Roll (ed.) *The Mixed Economy*, London: Macmillan.

Simon, H.A. (1977) *The New Science of Management Decision*, Englewood Cliffs, NJ: Prentice-Hall.

Smith, A. (1977) *The Wealth of Nations*, London: Methuen.

Stigler, G.J. (1968) *The Organization of Industry*, Homewood, Ill.: Richard D. Irwin.

Takezawa, S. and Whitehill, A.M. (1981) *Work Ways: Japan and America*, Tokyo: Japan Institute of Labour.

Trevelyan, G.M. (1978) *English Social History*, Harlow: Longman.

Tsuru, S. (1968) *Essays on Economic Development*, Tokyo: Kinokuniya.

UK Iron and Steel Statistics Bureau (annual) *Annual Statistics*, Croydon: Iron and Steel Statistics Bureau (1960–92).

Vaizey, J. (1974) *The History of British Steel*, London: Weidenfeld and Nicolson.

Vogel, E.F. (1979) *Japan as Number One: Lessons for America*, Cambridge, Mass.: Harvard University Press.

Webb, S. and Webb, B. (1920) *Industrial Democracy*, London: Longman.

Wedderburn, D. and Crompton, R. (1973) *Workers' Attitude and Technology*, Cambridge: Cambridge University Press.

White, M. and Trevor, M. (1983) *Under Japanese Management*, London: Heinemann Educational.

Wickens, P. (1987) *The Road to Nissan*, London: Routledge.

Williamson, O.E. (1975) *Markets and Hierarchies*, New York: Free Press.

Woodward, J. (1965) *Industrial Organisation: Theory and Practice*, London: Oxford University Press.

Yoshino, M.Y. (1968) *Japan's Managerial System*, Cambridge, Mass.: MITI Press.

Index

CPSIA information can be obtained
at www.ICGtesting.com
Printed in the USA
LVHW102244190522
719269LV00002B/13